Chemistry of Essential Oils and Food Flavours

Chemistry of Essential Oils and Food Flavours

Editors

Ian Southwell

Oscar Núñez

MDPI • Basel • Beijing • Wuhan • Barcelona • Belgrade • Manchester • Tokyo • Cluj • Tianjin

MDPI

Editors
Ian Southwell
Southern Cross University
Australia

Oscar Núñez
University of Barcelona
Spain

Editorial Office
MDPI
St. Alban-Anlage 66
4052 Basel, Switzerland

This is a reprint of articles from the Special Issue published online in the open access journal *Foods* (ISSN 2304-8158) (available at: https://www.mdpi.com/journal/foods/special_issues/essential_oil_food).

For citation purposes, cite each article independently as indicated on the article page online and as indicated below:

LastName, A.A.; LastName, B.B.; LastName, C.C. Article Title. *Journal Name* **Year**, *Volume Number*, Page Range.

ISBN 978-3-0365-5871-4 (Hbk)
ISBN 978-3-0365-5872-1 (PDF)

Contents

About the Editors

Ian Southwell

Ian Athol Southwell (Adjunct Professor in Plant Science at Southern Cross University, Lismore, Australia) completed a B.Sc. (Hons.) degree in 1967, an M.Sc. at the University of Sydney, Australia, and then a Ph.D. in 1982 at the University of Manchester Institute of Science and Technology. He has been engaged in plant chemistry research and development for more than 50 years, specialising in Australian essential-oil-bearing plants. Most of this research was conducted as a government research scientist in Australia, except for his PhD and postdoctoral appointments in Manchester, UK, and Indiana, USA. Ian has co-authored some 200 research papers, conference presentations, industry reports and book chapters, including co-editing "Tea Tree, the genus *Melaleuca*" for the "Medicinal and Aromatic Plants Industrial Profiles" series, in which he also co-edited the *Eucalyptus* Chemistry chapter for their *Eucalyptus* volume. He contributes to the international industry by assisting with overseas aid programmes, speaking at workshops and conferences and sitting on a number of essential oil committees and boards including the ISO Essential Oil Committee. At present, he is an Adjunct Professor in Plant Science at Australia's Southern Cross University.

Oscar Núñez

Oscar Núñez (Full Professor in Analytical Chemistry) studied chemistry at the University of Barcelona, where he also received his PhD in 2004. He worked as visiting researcher for half a year in the development of online pre-concentration methods using micellar electrokinetic chromatography (MEKC) at the University of Hyogo (Japan) in collaboration with Professor Terabe (father of the MEKC technique). Starting in October 2005, he joined the Kyoto Institute of Technology (Japan) as a two-year post-doc researcher working with Professor Tanaka in developing monolithic silica capillary columns under a fellowship from the Japan Society for the Promotion of Science. In November 2007, he joined the University of Barcelona again, and since June 2021, he has been a Full Professor in Analytical Chemistry (Department of Chemical Engineering and Analytical Chemistry, University of Barcelona). He has more than 150 scientific papers and book chapters to his name, and he is editor of six books on chromatographic and related separation techniques, LC-MS/MS, and sample preparation techniques in food and environmental analysis and capillary electrophoresis. He has participated, with more than 200 contributions, in international and national scientific congresses. He has extensive experience in the development of liquid chromatography methods with ultraviolet and fluorescence detection, liquid chromatography coupled to low- and high-resolution mass spectrometry, as well as in sample treatment procedures, for environmental and food analysis. His major research areas nowadays involve the development of analytical methodologies in combination with chemometric techniques for the characterization, classification, and authentication of food and natural products, as well as the prevention of food frauds....

Preface to "Chemistry of Essential Oils and Food Flavours"

Essential oils have important functions in nature. In addition to serving important functions in nature, essential oils are used commercially as in flavours, fragrances and healthcare products. Their properties present a challenge with respect to investigating their chemical structure, function, bioactivity, analysis and commercialization as value-added products. Analytical techniques for the investigation of essential oil chemistry are forever changing. The sophistication, adulteration or imitation of natural products is becoming difficult to detect, as the perpetrators of such adulterations also use modern techniques. Consequently, an understanding of the chemistry of essential oils that keeps abreast with the latest in instrumental and computational developments is paramount.

Thus, the objective of this Special Issue is to broadcast the latest advances in essential oil discoveries with respect to oil chemistry, methodology, instrumentation, bioactivity, chemical ecology, biosynthesis and authentication, especially in relation to foods.

Ian Southwell and Oscar Núñez

Editors

foods MDPI

Editorial

Editorial for the Special Issue, "Chemistry of Essential Oils and Food Flavours"

Ian Southwell [1,*] and Oscar Núñez [2,3]

1 Southern Cross Plant Science, Faculty of Science and Engineering, Southern Cross University, Lismore, NSW 2480, Australia
2 Department of Chemical Engineering and Analytical Chemistry, Faculty of Chemistry, University of Barcelona, Martí i Franquès 1-11, E-08028 Barcelona, Spain; oscar.nunez@ub.edu
3 Research Institute in Food Nutrition and Food Safety, University of Barcelona, Recinte Torribera, Av. Prat de la Riba 171, Edifici de Recerca (Gaudí), Santa Coloma de Gramenet, E-08921 Barcelona, Spain
* Correspondence: ian.southwell@scu.edu.au

Citation: Southwell, I.; Núñez, O. Editorial for the Special Issue, "Chemistry of Essential Oils and Food Flavours". *Foods* **2022**, *11*, 2182. https://doi.org/10.3390/foods11152182

Received: 13 July 2022
Accepted: 19 July 2022
Published: 22 July 2022

Publisher's Note: MDPI stays neutral with regard to jurisdictional claims in published maps and institutional affiliations.

Essential oils have important functions in nature. In addition, they are used commercially in categories such as flavors, fragrances, and health-care products. Their properties present a challenge when investigating their chemical composition, function, bioactivity, analysis, and commercialization as value-added products. Analytical techniques for the investigation of essential oil chemistry are continuously evolving. The sophistication, adulteration, or imitation of natural products is becoming difficult to detect as the perpetrators employ modern techniques. Consequently, an understanding of the chemistry of essential oils that keeps abreast with current instrumental and computational developments is paramount.

Thus, the objective of this Special Issue is to broadcast some of the latest progress in essential oil discoveries concerning oil chemistry, methodology, instrumentation, bioactivity, chemical ecology, biosynthesis, and authentication, especially regarding foods.

This Special Issue will be useful for all readers in terms of the novel information it provides on the chemistry of essential oils and food flavors.

The special issue includes five Research Articles and two Reviews, which are briefly described as follows:

(1) *Rosmarinus officinalis* L. (also known as *Salvia rosmarinus* Schleid) and *Lavandula angustifolia* Mill. belong to the Lamiaceae family, which comprises different genera whose biological activities are used in traditional medicines worldwide. *R. officinalis* has many different uses; its volatile essential oil (EO) and leaf extracts possess extensively studied biological properties, such as antioxidant, anti-inflammatory, antiproliferative, anticancer, and antiviral, among others. *R. officinalis* was investigated for its curative properties against ailments caused by biochemical, chemical, or biological agents. Garzoli et al. [1], in the study "Headspace/GC–MS Analysis and Investigation of Antibacterial, Antioxidant and Cytotoxic Activity of Essential Oils and Hydrolates from *Rosmarinus officinalis* L. and *Lavandula angustifolia* Miller", described the chemical composition and biological activities of these essential oils and their hydrolates. They are rich in bioactive molecules and have potential applications in various fields, including foods and beverages. Although the oils of these species have been thoroughly examined over the years, less is known about their headspace/GC–MS analysis, especially regarding hydrolates.

(2) Furocoumarins (FCs) are secondary plant metabolites produced in response to damage by pests and other stressful challenges. Among Rutaceae species, FCs are characteristic compounds of citrus peel. FCs are found in citrus fruits together with polymethoxyflavones (PMFs) and coumarins (Cs); the latter are also contained in several spices, particularly cinnamon. These chemicals have been extensively investigated for

their wide range of biological activities, making them interesting nutraceuticals for dietary supplements. Arigò et al. [2], in the study "Dietary Intake of Coumarins and Furocoumarins through Citrus Beverages: A Detailed Estimation by an HPLC-MS/MS Method Combined with the Linear Retention Index System", quantified 35 oxygen heterocyclic compounds (FCs, PMFs, and Cs) in various drinks to obtain informative data useful for the regulatory authorities for the establishment of official limits.

(3) Terpenes are naturally occurring compounds produced by plants that are of great commercial interest in the food, agricultural, cosmetic, and pharmaceutical industries due to their broad spectra of antibacterial, antifungal, anthelmintic, membrane permeation enhancement, and antioxidant biological activities. However, their applications are often limited by their volatility and the requirement for surfactants to produce stable and soluble products. Soto et al. [3], in the study "Yeast Particle Encapsulation of Scaffolded Terpene Compounds for Controlled Terpene Release", reported the development of a second generation of yeast particle-encapsulated terpene technology that incorporates the stimuli-responsive control of terpene release. Hence the use of biodegradable pro-terpene compounds enabled higher encapsulation stability whilst retaining the full biological activity of the parent terpene compound.

(4) Lemon essential oil (LEO) is a well-known flavoring agent with versatile biological activities. The main components of LEO are represented by monoterpenoids, being a complex mixture of limonene, γ-terpinene, citral, linalool, β-caryophyllene, α-pinene, and β-pinene, exhibiting anti-inflammatory and antioxidant effects. Pucci et al. [4], in the study "Biological Properties of a Citral-Enriched Fraction of Citrus *limon* Essential Oil", isolated and characterized four citral-enriched fractions of winter LEO. Overall, the reported results encourage the application of EO fractions, enriched in citral, in the nutraceutical industry, not only for their organoleptic properties but also for their protective action against inflammation and oxidative stress.

(5) Although essential oils have a long history of use in medicinal plant, fragrance, and flavour applications, their potential human application is only now being explored. For example, Mastiha essential oil, rich in myrcene, α-pinene, and β-pinene, is obtained from water distillation of the resin of *Pistacia lentiscus* and has a commercial value not only as an aromatic resin but also as a medicinal plant [5]. Papada et al. [6] in the study "An Absorption and Plasma Kinetics Study of Monoterpenes Present in Mastiha Oil in Humans" identify and quantify these bioactive monoterpenes in the plasma following the acute consumption of this Mediterranean essential oil. This is the first study showing the bioavailability of this oil in the plasma as soon as 30 min after consumption.

(6) Conventional pesticides present several problems. An increasing global population has increased the use of pesticides in agriculture [7]. This overuse has led to serious health and ecological issues. Also, the problem is compounded by increasing insecticide and fungicide resistance. There is a movement to use more 'nature-friendly' bioactive compounds present in natural products. Plant-derived essential oils with insecticidal, fungicidal, and bacteriocidal activity fall into this category. Certain essential oils can enhance the action of commercially available pesticides. Dassanayake et al. [8] have completed a comprehensive review entitled "Synergistic Field Crop Pest Management Properties of Plant-Derived Essential Oils in Combination with Synthetic Pesticides and Bioactive Molecules: A Review". These authors review: the historical background and development of natural products in agriculture; the sources and chemical composition of plant-derived essential oils; pesticidal and fungicidal action mechanisms; synergistic and hybridized insect pest management products; synergistic and hybridized fungicidal activity; and novel developments.

(7) Lemon oils are one of the most prominent essential oils traded and consumed today. Thus far, knowledge of a more obscure source, *Backhousia citriodora* (BC), has been limited. Southwell [9] presents a timely and comprehensive review entitled "Backhousia citriodora F. Muell. (Lemon Myrtle), an Unrivalled Source of Citral." In this allegedly

unrivalled source of lemon constituents, details of the taxonomy, etymology, habit, distribution, chemotypes, agronomy, uses, essential oil, oil chemistry, plant bioactivity, toxicology, standards, and commerce are examined. BC is compared with other sources of lemon constituents to determine its superiority. The plant's applications for flavour, fragrance, and health care are highlighted. The work also studies the plant's chemistry and its use for determining authenticity via Australian and International standardisation [10,11]. Health-care investigations on both the oil and extract are outlined, and its safety and toxicology (e.g. GRAS status) are highlighted. It also attempts to summarise the current commercial situation regarding production, value, and markets [12].

Author Contributions: I.S. and O.N., writing. All authors have read and agreed to the published version of the manuscript.

Conflicts of Interest: The authors declare no conflict of interest.

References

1. Garzoli, S.; Masci, V.L.; Franceschi, S.; Tiezzi, A.; Giacomello, P.; Ovidi, E. Headspace/GC–MS Analysis and Investigation of Antibacterial, Antioxidant and Cytotoxic Activity of Essential Oils and Hydrolates from *Rosmarinus officinalis* L. and *Lavandula angustifolia* Miller. *Foods* **2021**, *10*, 1768. [CrossRef] [PubMed]
2. Arigò, A.; Rigano, F.; Russo, M.; Trovato, E.; Dugo, P.; Mondello, L. Dietary intake of coumarins and furocoumarins through citrus beverages: A detailed estimation by a HPLC-MS/MS method combined with the linear retention index system. *Foods* **2021**, *10*, 1533. [CrossRef] [PubMed]
3. Soto, E.R.; Rus, F.; Li, H.; Garceau, C.; Chicca, J.; Elfawal, M.; Gazzola, D.; Nielsen, M.K.; Urban, J.F.; Aroian, R.V.; et al. Yeast particle encapsulation of scaffolded terpene compounds for controlled terpene release. *Foods* **2021**, *10*, 1207. [CrossRef] [PubMed]
4. Pucci, M.; Raimondo, S.; Zichittella, C.; Tinnirello, V.; Corleone, V.; Aiello, G.; Moschetti, M.; Conigliaro, A.; Fontana, S.; Alessandro, R. Biological properties of a citral-enriched fraction of citrus limon essential oil. *Foods* **2020**, *9*, 1290. [CrossRef] [PubMed]
5. Paraschos, S.; Magiatis, P.; Mitakou, S.; Petraki, K.; Kalliaropoulos, A.; Maragkoudakis, P.; Mentis, A.; Sgouras, D.; Skaltsounis, A.-L. In Vitro and In Vivo Activities of Chios Mastic Gum Extracts and Constituents against Helicobacter pylori. *Antimicrob. Agents Chemother.* **2006**, *51*, 551–559. [CrossRef] [PubMed]
6. Papada, E.; Gioxari, A.; Amerikanou, C.; Galanis, N.; Kaliora, A.C. An Absorption and Plasma Kinetics Study of Monoterpenes Present in Mastiha Oil in Humans. *Foods* **2020**, *9*, 1019. [CrossRef] [PubMed]
7. Popp, J.; Pető, K.; Nagy, J. Pesticide productivity and food security. A review. *Agron. Sustain. Dev.* **2012**, *33*, 243–255. [CrossRef]
8. Dassanayake, M.K.; Chong, C.H.; Khoo, T.-J.; Figiel, A.; Szumny, A.; Choo, C.M. Synergistic Field Crop Pest Management Properties of Plant-Derived Essential Oils in Combination with Synthetic Pesticides and Bioactive Molecules: A Review. *Foods* **2021**, *10*, 2016. [CrossRef] [PubMed]
9. Southwell, I.A. *Backhousia citriodora* F. Muell. (Lemon Myrtle), an Unrivalled Source of Citral. *Foods* **2021**, *10*, 1596. [CrossRef] [PubMed]
10. *AS 4941-2001*; Oil of *Backhousia citriodora*, Citral Type (Lemon Myrtle Oil). Standards Australia: Sydney, Switzerland, 2001.
11. *ISO/DIS 5093-2021*; Essential Oil of Lemon Myrtle (*Backhousia citriodora* F. Muell.), Citral Type. International Standards Organisation: Geneva, Switzerland, 2021.
12. Australian Native Food & Botanicals. Lemon Myrtle. Available online: https://anfab.org.au/main.asp?_=Lemon%20Myrtle (accessed on 4 July 2022).

foods

MDPI

Review

Synergistic Field Crop Pest Management Properties of Plant-Derived Essential Oils in Combination with Synthetic Pesticides and Bioactive Molecules: A Review

Mackingsley Kushan Dassanayake [1], Chien Hwa Chong [2,*], Teng-Jin Khoo [1], Adam Figiel [3], Antoni Szumny [4] and Chee Ming Choo [5]

1 School of Pharmacy, Faculty of Science and Engineering, University of Nottingham Malaysia, Jalan Broga, Semenyih 43500, Malaysia; hyxmd1@nottingham.edu.my (M.K.D.); tengjin.khoo@nottingham.edu.my (T.-J.K.)
2 Department of Chemical and Environmental Engineering, Faculty of Science and Engineering, University of Nottingham, Jalan Broga, Semenyih 43500, Malaysia
3 Institute of Agricultural Engineering, Wrocław University of Environmental and Life Sciences, Chełmońskiego 37a, 51-630 Wrocław, Poland; adam.figiel@upwr.edu.pl
4 Department of Chemistry, Wrocław University of Environmental and Life Sciences, Norwida 25, 50-375 Wrocław, Poland; antoni.szumny@upwr.edu.pl
5 Centre for Water Research, Faculty of Engineering, Built Environment and Information Technology, SEGi University Kota Damansara, Petaling Jaya 47810, Malaysia; choocheeming@segi.edu.my
* Correspondence: ChienHwa.Chong@nottingham.edu.my

Abstract: The management of insect pests and fungal diseases that cause damage to crops has become challenging due to the rise of pesticide and fungicide resistance. The recent developments in studies related to plant-derived essential oil products has led to the discovery of a range of phytochemicals with the potential to combat pesticide and fungicide resistance. This review paper summarizes and interprets the findings of experimental work based on plant-based essential oils in combination with existing pesticidal and fungicidal agents and novel bioactive natural and synthetic molecules against the insect pests and fungi responsible for the damage of crops. The insect mortality rate and fractional inhibitory concentration were used to evaluate the insecticidal and fungicidal activities of essential oil synergists against crop-associated pests. A number of studies have revealed that plant-derived essential oils are capable of enhancing the insect mortality rate and reducing the minimum inhibitory concentration of commercially available pesticides, fungicides and other bioactive molecules. Considering these facts, plant-derived essential oils represent a valuable and novel source of bioactive compounds with potent synergism to modulate crop-associated insect pests and phytopathogenic fungi.

Keywords: phytochemicals; synergism; essential oils; fractional inhibitory concentration; insect mortality rate; phytopathogenic fungi; insect pests; pesticide resistance; fungicide resistance

Citation: Dassanayake, M.K.; Chong, C.H.; Khoo, T.-J.; Figiel, A.; Szumny, A.; Choo, C.M. Synergistic Field Crop Pest Management Properties of Plant-Derived Essential Oils in Combination with Synthetic Pesticides and Bioactive Molecules: A Review. *Foods* **2021**, *10*, 2016. https://doi.org/10.3390/foods10092016

Academic Editors: Ian Southwell and Oscar Núñez

Received: 24 July 2021
Accepted: 23 August 2021
Published: 27 August 2021

Publisher's Note: MDPI stays neutral with regard to jurisdictional claims in published maps and institutional affiliations.

1. Introduction

The demand for the production of crops is rising due to the increasing global population, which may exceed 35% by 2050 [1]. This has led to a 15–20-fold use of pesticides in order to enhance the availability of crop yields across the globe [2]. Pesticides are chemical agents that are either synthetically made or naturally occurring, which can be classified as insecticides, fungicides, herbicides, nematicides, rodenticides, etc. Approximately, 2 million metric tons of pesticides are used in agriculture across the globe annually, where countries like China, the USA and Argentina are the major contributors towards pesticide use, and it has been estimated that annual pesticide usage will soon increase up to 3.5 million metric tons worldwide [3]. It has been reported that around 47.5% of herbicides, 29.5% of insecticides, 17.5% of fungicides and the remaining 5.5% of other pest management agents account for all pesticides used worldwide [4]. However, the overuse of synthetic pesticides has led to serious health and ecological hazards, such as the increased risk of cancers, as well as cardiovascular, neurological, endocrine-related health issues and the potential damage done to non-target animals and plants that exist within the parameters of the agent applied [5]. For example, workers who were handling pesticides that consist of hexachlorocyclohexane (HCH) have experienced neurological symptoms. It was reported in 1992 by the National Institute of Occupational Health (NIOH) that paddy field workmen who were spraying insecticides containing methomyl showed abnormalities in their ECG, serum LDH and cholinesterase levels [6]. Chlorpyrifos is one of the most widely used synthetic pesticides in the history of agricultural practices, and the application of this agent can contaminate the soil and groundwater and known to be highly toxic to aquatic life [7]. The environmental, ecotoxicological and health consequences of the widespread application of synthetically made chemical pesticides and fungicides, as well as the development of resistance to these agents, have resulted in a heightened concern and interest among researchers and consumers to focus more on natural and sustainable products with fewer synthetic pesticides, insecticides, fungicides and herbicides [5].The quality of nutrition, food security and sustainability have become very important agenda issues in Sustainable Development Goal 2 (SDG2) established by the United Nations in 2015, and according to current estimates of SDG2, about 8.9% or 690 million people of the world population are in hunger; thus, it may not be possible to achieve zero hunger by the year 2030 [8]. Hence, the United Nations World Food Program aims to alleviate worldwide starvation by the year 2050. There exists a potential to integrate essential oils (EOs) and bioactive compounds from plants, herbs, fruit waste and enzymes of ripening fruits into agricultural practices. Essential oils (EOs) and bioactive compounds from plants, herbs, fruit waste and enzymes of fruits or biomaterials are potential crop protection agents [9]. Essential oils are odoriferous volatile natural oils that can be characterized by their aromatic and lipophilic nature [10]. These EOs are promising sources of naturally occurring bioactive compounds that show pesticidal and fungicidal activities [11]. Plants produce both primary (e.g., sugars and acids) and secondary metabolites, where EOs are largely composed of bioactive secondary metabolites like monoterpenes, esters, sesquiterpenes, phenols, aldehydes, oxides and ketones that are synthesized both internally and externally by plants [12,13]. Essential oils are abundantly found in aromatic plants, where more than 3000 types of EOs have been identified and about 300 essential oil variants have been commercialized [10,11,14,15]. Families of plants that are frequently studied for their essential oils include *Lauraceae*, *Myrtaceae*, *Lamiaceae*, *Rutaceae*, *Apiaceae*, *Asteraceae*, *Poaceae*, *Cupressaceae*, *Piperaceae* and *Zingiberaceae* [16–18]. Nonetheless, the demand for novel pesticidal and fungicidal products from natural sources is increasing, and it has been estimated that around 40%–50% of the crop yields of maize, barley, wheat, rice, potatoes, sugar beets and soybeans harvested worldwide are dissipated each year, largely due to pesticide resistance in crop-consuming insects [2]. The registration process for a new fungicide or pesticide usually requires the registrant (e.g., manufacturer) to analyze and conduct different laboratory-based tests [19]. These tests will define the chemistry of the new fungicide of pesticide, as well as the potential hazards to humans, domestic animals, and the proximal environmental and the impact on non-target organisms. Data that include the identity, chemical and physical properties of the active ingredient present in the product, as well as analytical methods, the proposed label and uses, human and environmental toxicity, safety data sheets, efficacy associated with the intended use, container management, residues resulting from the pesticide product usage and the disposal of product waste, are needed to support the application of a pesticide or fungicide registration during its full life-cycle [19,20]. The generation and verification of such data for a single compound may take many years and can be expensive [21]. Hence, there is a growing interest and continuous demand to discover new insecticidal, fungicidal and herbicidal agents with novel mechanisms of action, accompanied by efforts to ensure safety and reduce production cost.

Currently, research has been implemented on various chemical properties and biological activities like antioxidant, anticancer, antimicrobial, antiviral and pesticidal effects of plant-derived essential oils [22]. The following review paper emphasizes the impact of potent plant-derived essential oils and their bioactive compounds that synergistically integrate with synthetic pesticides and other novel molecules for crop preservation.

2. Historical Background and Development of Natural Products in Agriculture

Bioactive compounds present in these natural products can be applied as pesticidal, insecticidal and fungicidal agents [23]. The origins of many synthetic pesticidal, insecticidal and antifungal agents can be traced back from a variety of natural products since the introduction and commercialization of penicillin [24–26]. The use of plant-based pesticidal agents has been reported since ancient times, where extracts of poisonous herbs were used to control crop-consuming insect pests about 4000 years ago [27]. Nicotine sulfate, extracted from the leaves of tobacco plants, was applied as a natural insecticide in the seventeenth century, and compounds like pyrethrum derived from chrysanthemums flowers and rotenone extracted from the roots of tropical vegetables were used as natural pesticides in the nineteenth century [28]. The use of naturally occurring substances as fungicidal agents has been reported since the seventeenth century, when sea salt and lime were used to treat wheat in order to prevent the growth of bunt caused by fungi [29]. Another important discovery was made by the French botanist Pierre-Marie-Alexis Millardet, who concluded that copper sulfate, which is a naturally occurring substance, was able to effectively control and reduce downy mildew of certain fruits like grapes [30]. Natural products and their bioactive derivatives constituted about 36% of ingredients present in commercially available pesticides from 1997 to 2010. For example, soil-borne bacteria and *Streptomyces avermitilis* and *Saccharopolyspora spinosa* were used to produce natural pesticides known as avermectin and spinosyn, which can effectively cause the paralysis of insect pests [31]. Avermectin is an award-winning natural pesticidal agent that was isolated from the actinomycete species of bacteria known as *S. avermitilis*. Glufosinate, also known as phosphinothricin, is a naturally occurring broad-spectrum herbicidal agent produced by the bacteria of *Streptomyces* spp. [23]. This bacterial-derived compound was commercialized as an herbicide by the German pharmaceutical company named Bayer under the trade name of Finale [32,33]. The herbicidal action of glufosinate works by inhibiting the enzyme glutamine synthetase, resulting in the buildup of ammonia in the thylakoid lumen of plants and leading to photophosphorylation decoupling. The British pharmaceutical company named Corteva Agriscience commercialized a fungicide known as fenpicoxamid that was derived from antimycin, which is naturally produced by *Streptomyces* spp. bacteria. Fenpicoxamid works by inhibiting cellular respiration in fungi. The annual gross of fenpicoxamid and glufosinate exceeded USD 1 billion after introducing them to the market. Other examples of herbicides include the *Streptomyces* spp. produced tentoxin and the fungal *Alternaria alternate* (Fries)-derived thaxtomin [23]. These herbicidal agents were able to disrupt energy metabolism cellulose biosynthesis. Cornexistin is a fungal metabolite derived from *Paecilomyces variotii*, which acts as a broad-spectrum herbicidal agent against maize via the inactivation of enzymes known as aminotransferases [23,34].

3. Sources and Chemical Composition of Plant-Derived Essential Oils

Several species of plants consist of volatile essential oils, in which different plant parts like leaves, barks, peels, flowers, seeds, buds and roots can be diverse sources of various essential oils [35]. Plant-based essential oils are complex mixtures of naturally occurring polar and nonpolar compounds [36]. These essential oils have been classified into four primary groups as terpenes, derivatives of benzene, hydrocarbons and other forms of miscellaneous aromatic compounds [37,38]. Terpenes like monoterpenes and monoterpenoids are the most abundant and major representative molecules that constitute about 90% of EOs [39]. Plant-derived EOs are largely composed of carbon hydrocarbons including the following: acyclic alcohols like geraniol, linalool and citronellol; cyclic alcohols like terpeniol, menthol and isopulegol; bicyclic alcohol compounds like verbenol and borneol; phenols that include carvacrol and thymol; ketones like menthone, carvone and thujone; aldehydes that include citral and citronellal;acids like chrysanthemic acid; and oxides like cineole [35]. Terpenes present in these EOs are further classified into the following groups according to their molecular weight: hemiterpenes (C5), monoterpenes (C10), sesquiterpenes (C15), diterpenes (C20), triterpenes (C30) and tetraterpenes (C40). Aromatic compounds occur less frequently compared to terpenes and are natural derivatives of phenylpropane compounds like cinnamaldehyde, aldehyde, cinnamic alcohol, as well as phenols that include eugenol and chavicol, methoxy derivatives like elemicine, methyl eugenols, anethole, estragole and methylenedioxy compounds like myristicine, apiole and safrole [14]. Although EOs are present in a variety of plants, their extraction and productivity are relatively time consuming and expensive processes, since very small amounts of pure EOs can be harnessed from a large amount of raw plant material [35,40].

4. Pesticidal and Fungicidal Action Mechanisms of Plant-Derived Essential Oils

Plant-derived essential oils consist of intrinsic properties that can interfere with biochemical, physiological and metabolic functions of insects and fungi by altering the biological activities of target sites of these organisms [41,42]. Anti-insect pest and antifungal agents from botanical EOs can have either narrow-spectrum or broad-spectrum activity, in which narrow-spectrum agents will only affect a particular species of insects or fungi and broad-spectrum agents are effective against a wide range of fungi or insect pests [43]. Additionally, these botanical agents can be classified as fungistatic, which only slow down the growth and multiplication of fungi but do not actually kill them, or asfungicidals that directly promote the cellular destruction of fungal organisms [44]. In case of anti-insect pest plant-derived agents, these can be classified as insect repellents, which consist of chemical properties that can simply repel insects, or as insecticides, which are lethal to insects and cause mortality upon contact [45].

4.1. Mode of Action of Insecticidal Essential Oils

Several molecular studies have revealed the action mechanisms of plant-derived essential oils that show the efficiency of pesticidal and insect repellent activity. The EO metabolite-mediated inhibition of acetylcholinesterase (AChE) and octopamine pathways of insects [46–50] (Figure 1) has been well investigated and documented. Among these mechanisms, the inhibition of AChE is one of the most exploited, since AChE is an enzyme that plays a crucial role in and neuromuscular and neuronal communication in insects [51–53]. AChE inhibition can cause neurotransmitter toxic effect on insect pests by the membrane disruption of the postsynaptic junction that leads to the interference of nerve current [54–56]. Octopamine is an important hormone associated with the nervous system of insects [57]. This neurohormone is present as octopamine-1 and octopamine-2 and respectively functions as a neurotransmitter and as a neuromodulator in insects, in which the inhibition of octopamine will cause the impairment of physiological modulation associated with muscle juncture and homeostasis of insect bodily fluids, which can alter their octopamine-mediated nervous system [58–63]. Plant-derived EOs are also capable of inhibiting GABA receptors present in insects, which can suspend GABA from binding with GABArs (GABA receptors) in extrasynaptic synaptic membranes [64–66] (Figure 1). Furthermore, phytochemical metabolites from plant-derived EOs can inhibit or interfere the activities of enzymes associated with the metabolism of xenobiotics and respiration of insects like CarEs, chitin, cytochrome P450s, ATP-binding cassette transporters and GSTs [67–69].

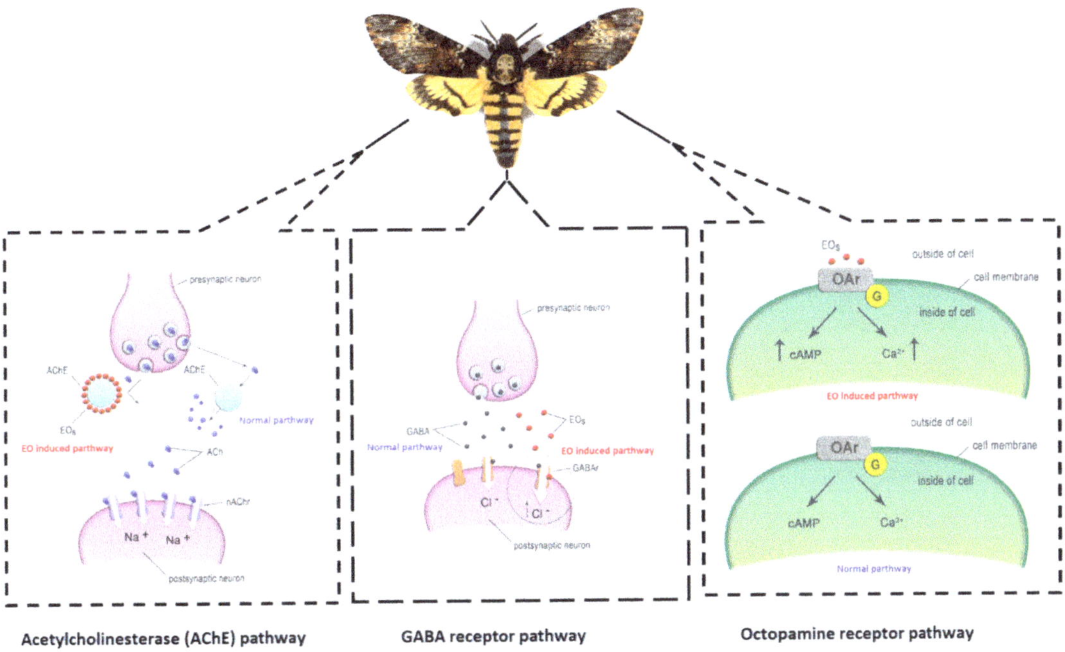

Figure 1. Insecticidal action mechanisms of plant-derived essential oils.

4.2. Mode of Action of Fungicidal Essential Oils

Plant-derived essential oils have multiple mechanisms of action to inhibit the growth and activity of fungi. Target sites of these EO metabolites include the biosynthesis of cell wall, ATPases activity, efflux pumps, quorum sensing/biofilm formation and cell membrane structure and integrity in fungi [70–74] (Figure 2). Essential oils that disrupt cell wall biosynthesis work by inhibiting the formation of components like chitin and β-glucans, which are necessary for the synthesis of fungal cell walls [75]. Ergosterol is an essential compound associated with fungal cell membranes and their biosynthetic pathways. The inhibition of ergosterol by EOs will cause structural, metabolic and osmotic instability in fungal cells, leading to compromised multiplication and virulence [76–79]. Certain EOs can affect the ATPases activity of fungi by interfering with the function enzymes associated with fungal mitochondria. The inhibition of mitochondrial enzymes like malate dehydrogenase, succinate dehydrogenase and lactate dehydrogenase can alter the level of reactive oxygen species and ATP, which leads to the diminishing of mitochondrial content that is essential for fungal metabolic pathways [80]. Efflux pumps are proteinaceous transporters localized in the cell membranes of both prokaryotic and eukaryotic cells. In fungi, these are important structures that mediate nutrient uptake, medium acidification and antifungal resistance. These efflux pumps are target sites of certain metabolites associated with plant-derived essential oils in modifying or reversing antifungal resistance [80–82]. Plant-derived EOs are also capable of attenuating quorum-sensing (QS) activity in fungi, in which certain phytochemical metabolites present in these essential oils can inhibit cell-to-cell communicating QS signaling molecules like N-acyl homoserine lactones (AHLs), tyrosol, α-(1,3)-glucan and tryptophol, and fungal pheromones like a-factor and α-factor [83–88].

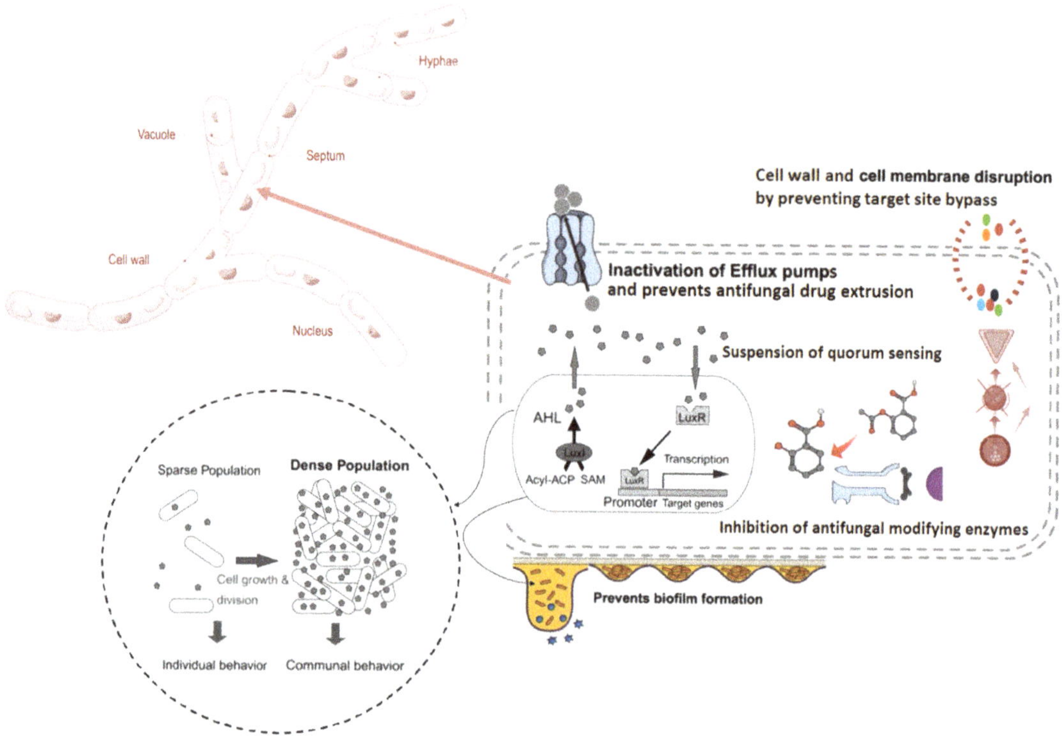

Figure 2. Antifungal action mechanisms of plant-derived essential oils.

5. Synergistic and Hybridized Insect Pest Management Products of Botanical Essential Oils

5.1. As Homosynergistic Agents

Plant-derived bioactive metabolites present in EOs are capable of interacting synergistically to increase pesticidal action. A study revealed that essential oil phytochemical compounds thymol and 1,8-cineole (Figure 3) interacted synergistically with pulegone to induce larvicidal activity against *Plutella xylostella* (Linnaeus) (diamondback moth). 1,8-cineole and pulegone (Figure 3) combination indicated the highest synergistic activity with a larval mortality rate of 90% in the study. The investigation further elucidated that thymol and 1,8-cineole were able to affect the levels of enzymes like carboxylesterase esterase, glutathione transferases and acetylcholinesterase associated with *P. xylostella* [89]. Rosemary essential oil compounds camphor (Figure 3) and 1,8-cineole indicated synergistic insecticidal action against the moth species known as *Trichoplusia ni* (cabbage looper). The study revealed that the mixture of these compounds (103 µg of 1,8-cineole and 150 µg of camphor) indicated a larval mortality rate >80% in both contact and fumigant assays with a penetration rate >40% in 60 min of application [90]. A similar study conducted by Tak and Isman [91] revealed that 1,8-cineole and camphor isolated from the essential oil of *Rosmarinus officinalis* were synergistically active when combined against *Trichoplusiani* (Hübner) larvae. A compound combination ratio of 60:40 of 1,8-cineole and camphor indicated a larvae mortality rate of 93.3 ± 6.7 in the study [91]. Binary mixtures of essential oil compounds α-terpineol (Figure 3) and thymol were able to synergize the biopesticidal activity of 1,8-cineole and linalool (Figure 3) against swinhoe larvae *Chilopartellus* (Swinhoe) at a dose of 189.7 µg [92]. An investigation conducted by Hummelbrunner and Isman [93] revealed that complex mixtures of *trans*-anethole, citronellal (Figure 3), α-terpineol and thymol were able to interact synergistically and mediate acute toxicity to *S. litura* Fab. (tobacco cutworms) when topically administered at a dose of 40.6 μg [93]. Liu et al. [94] indicated that essential oils extracted from *Cinnamomum camphora* (L.) Presl. seeds and *Artemisia princeps* Pamp leaves exhibited synergistic insecticidal and repellent activity against crop pests like *Sitophilusoryzae* L. (rice weevil) and *B. rugimanus* Bohem when combined at a concentration ratio of 1:1 [94]. A study showed that cinnamon oil was able to synergize the larvicidal activity of rotenone against *Spodoptera litura* (F.) at a mixture ratio of 1:35 and concentration of 506 mg/L within 72 h of exposure [95]. Essential oil compounds γ-terpinene and terpinen-4-ol (Figure 3) isolated from the extracts of *Majorana hortensis* Moench were able to synergistically mediate insecticidal activity against *Aphis fabae* (Scopoli) and *S. littoralis* [96]. Andrés et al. [97] showed that binary mixtures of essential oil compounds terpinolene and safrole (Figure 3) extracted from *Piper hispidinervum* were able to induce synergistic antifeedant effect on crop-related pests like *Leptinotarsa decemlineata* (Say), *S. littoralis*, *Rhopalosiphum padi*(Linnaeus) and *Myzuspersicae* (Sulzer) [97]. Furthermore, an investigation showed that a binary mixture composed of limonene and carvone (Figure 3) at a concentration ratio of 6:2 displayed synergistic pesticidal activity against *Tribolium castaneum* (Herbst) (red flour beetle) adults at 10.84 μg and larvae at 30.62 μg [98]. Examples of insecticidal homosynergistic plant-derived EOs and their compounds are summarized in Table 1.

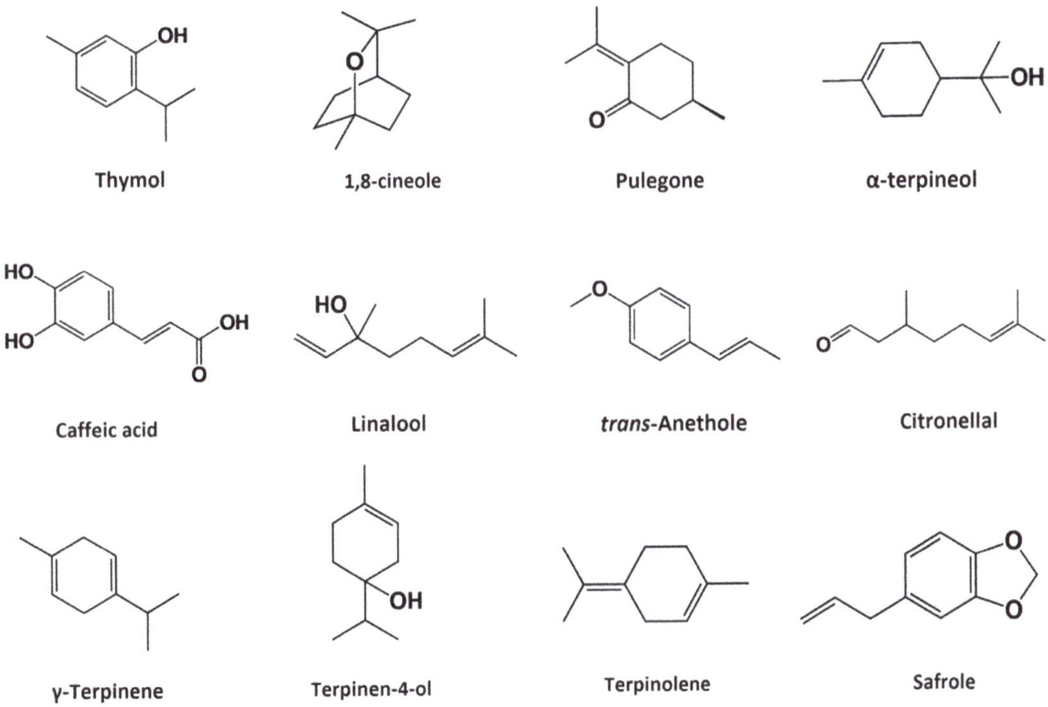

Figure 3. *Cont.*

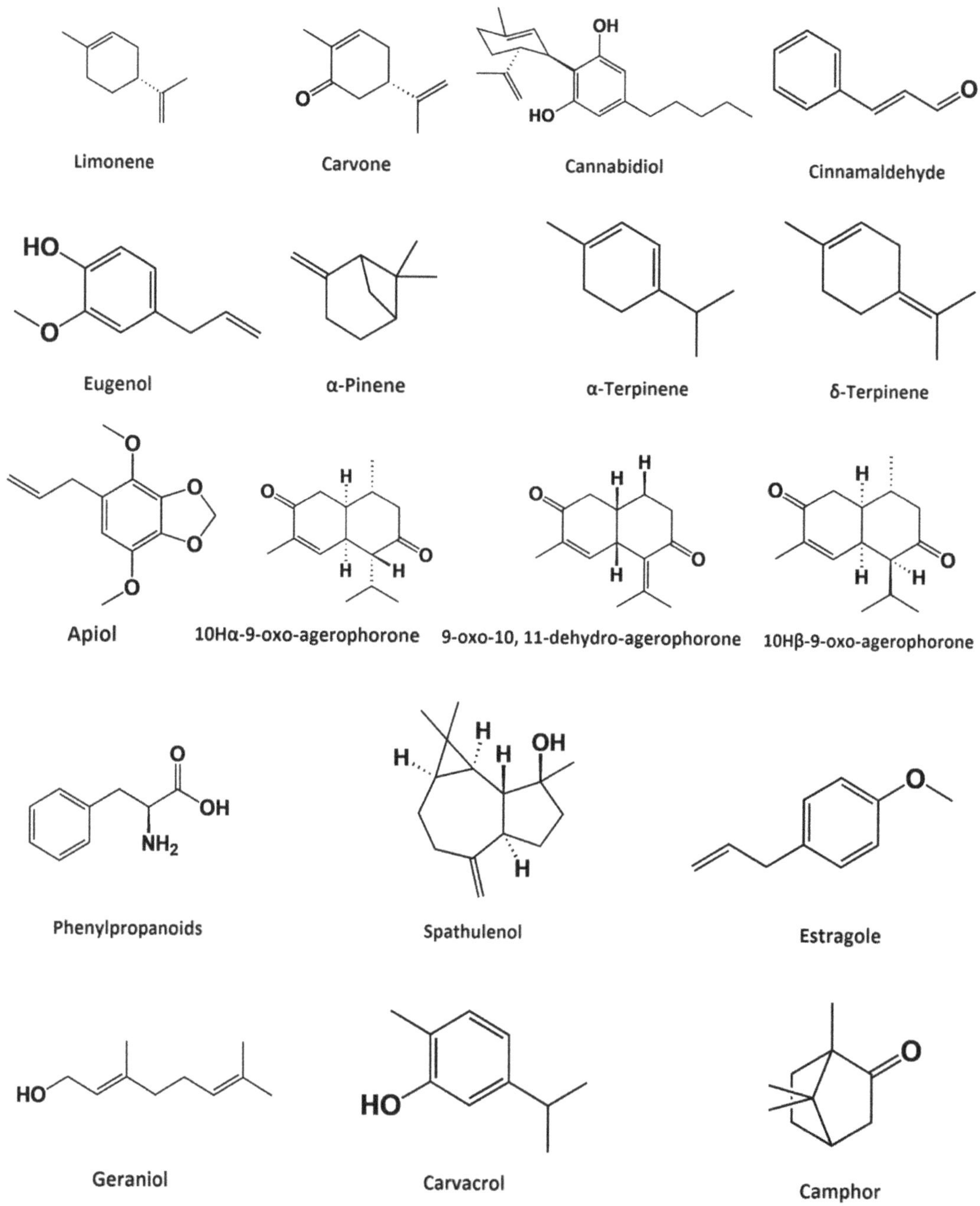

Figure 3. Phytochemical compounds isolated from plant-derived essential oil synergists.

5.2. As Enhancers of Commercial Insecticides

Certain essential oils and their representative phytochemical constituents are capable of enhancing the insecticidal action of commercially available synthetic chemical pesticides.

A study conducted by El-Meniawi et al. [99] showed that EOs from *Simmodsiachinesis*, *Allium sativum*, Fam. and *Mentha piperita* Fam. were able to synergistically enhance the activity of cyhalothrin, diuron and malathion, respectively, at concentrations ranging from 0.1 to 100 µm against *Bemisia tabaci* (Gennadius) (silver leaf whitefly). Further investigations in this study showed that these combinative agents induced the inhibition of the entomic enzymes ATPase, chitinase and acetylcholinesterase [99]. An investigation revealed the pesticide susceptibility of *Myzus persicae* (Sulzer) (green peach aphid) to imidacloprid and spirotetramat after individually combining them with *Thymus vulgaris* and *Lavandula angustifolia*, thymol and linalool, respectively. Imidacloprid with *L. angustifolia* combinative treatment indicated the highest synergism ratio of 19.8 in the study [100]. A similar study showed that rapeseed oil and soya oil enhanced the pesticidal action of pirimicarb and imidacloprid against *Myzuspersicae* (Sulzer) [101]. The essential oil compound linalool isolated from *Ocimumbasilicum* (Linnaeus) enhanced the pesticidal effect of deltamethrin against *Spodoptera frugiperda* (J.E. Smith) (all armyworm). The study showed that the dose of deltamethrin can be reduced by more than 6-fold by the application of 480 µg/µL of *O. basilicum* essential oil [102]. Another study indicated that deltamethrin at 9.62 µL and linalool at 0.177 µL combination induced enhanced insecticidal activity against *S. frugiperda* larvae, resulting in 95.75% mortality in 24 hours. The same study showed that linaloolatat0.177 µL enhanced the pesticidal activity of Decis® (25CE) at 0.25 µL, resulting in 100% larval mortality [103]. A recent research study conducted by Ismail (2021) showed that garlic oil was able to synergize and enhance the insecticidal action of chlorpyrifos and cypermethrin up to 9-fold against the crop pest *S. littoralis*. The study further elucidated that these combinative agents induced the inhibition of enzyme pathways associated with oxidase, glutathione S-transferase and general esterase (ά-β-EST) of the tested insect pest [104]. Mantzoukas et al. [105] stated that the cannabidiol (Figure 3) present in the essential oil of the Cannabis plant synergized the commercially available biopesticides madex, azatin and helicovex against the four crop pests *S. zeamais*, *Rhyzopertha dominica* (Fabricius), *Prostephanus truncates* (Horn) and *Trogodermagranarium* (Everts) at doses ranging from 500 to 3000 ppm [105]. Examples of commercially available synthetic pesticides used in combination with plant-derived essential oils and their compounds are summarized in Table 1.

6. Synergistic and Hybridized Fungicidal Activity of Botanical Essential Oils

6.1. As Homosynergistic Agents

Bioactive phytochemical metabolites present in EOs have been found to interact synergistically to mediate antifungal activity. A study revealed that EOs isolated from thyme, clove and lemongrass demonstrated high antifungal activity, which completely inhibited the growth of mycelium of *Fusarium oxysporum* (Sacc.) and *Fusarium circinatum* (Nirenberg and O'Donnell) at a concentration of 1000 µL/L [106]. Another study indicated that the essential oil combination of thyme, cinnamon, lime and clove induced antifungal activity against the crop-degrading fungus *Colletotrichum gloeosporioides* (Penz)and reduced the damage of crops [107]. Nardoni et al. [108] stated that EOs extracted from *Thymus vulgaris*, *Origanum vulgare*, *O. basilicum*, *Foeniculu mvulgare*, *Illicium verum*, *Syzygium aromaticum*, *Origanum majorana*, *Rosmarinus officinalis*, *Citrus sinensis*, *Citrus bergamia*, *Cymbopogon citrates*, *Salvia sclarea*, *Citrus aurantium*, *Citrus paradise* and *Citrus limon* showed synergistic antifungal activity against *P. funiculosum* and *M. racemosus* with a FICI of <0.5 for both fungi [108]. An investigation conducted by Bedoya-Serna et al. [109] showed that a nanoemulsion composed of a mixture of oregano and sunflower essential oil was synergistically active against *Fusarium* sp., *Cladosporium* sp. and *Penicillium* sp., which suspended their fungal spore formation at a concentration 0.1 mL [109]. Essential oils extracted from *Thymus vulgaris* and *O. vulgare* interacted synergistically to mediate antifungal activity against *Fusarium* spp. with FICIs ranging from 0.375 to 0.5 when used in combination. Moreover, the study showed that the best synergistic activity of the essential oil combination was demonstrated against *F. moniliforme* with a FICI of 0.375 at an indicative MIC and MFC of 0.156 μL/mL [110]. An investigation carried out by Yen and Chang [111] indicated that cinnamaldehyde and eugenol isolated from cinnamon essential oil were synergistically fungicidal against *L. sulphureus*. The study revealed that the MIC of the cinnamaldehyde and eugenol (Figure 3) combination was90% lower compared to their stand-alone treatments [111]. Hartati [112] stated that combining essential oils extracted from *Cymbopogon nardus* (citronella) and *Azadirachta indica* (neem) at a concentration ratio of 1:1 was synergistic and effective against the fungal pathogen of patchouli plants known as *Synchytriumpogostemonis* S.D.Patil and Mahab [112]. A study revealed that a combination of essential oils from *Syzygium aromaticum* (Linn.) (clove) and *Cinnamonum zeylanicum* (cinnamon) mediated synergistic fungicidal activity against a crop disease causing *Aspergillus niger*, *Alternaria alternate* (Fries) Keissler, *Colletotrichum gloeosporioides* (Penzig), *Lasiodiplodia theobromae* (Patouillard) Griffon and Maublanc, *Plasmopara viticola* (Berkeley and Curtis) and *Rhizopus stolonifer* (Ehrenberg) Vuillemin. The best synergistic antifungal activity was observed for clove oil and cinnamon oil (9:1) with a FICI of 0.55 against *P. viticola* in the study [113]. A research conducted by Yu et al. [114] indicated that essential oil compounds terpinolene, terpinen-4-ol, δ-terpinene, α-pinene, 1,8-cineole, α-terpineol and α-terpinene (Figure 3) isolated from *Melaleuca alternifolia* (tea tree) interacted synergistically to mediate antifungal activity against *Botrytis cinerea* (Persoon). According to the results of the study, the highest antifungal synergism was observed for terpinen-4-ol and α-terpineol combination (1:1 ratio), which indicated a mycelial growth inhibition rate of 99.46% ± 0.76%, and scanning electron microscopic analysis revealed that these compounds made pronounced alterations in the cell wall ultrastructure, mycelial morphology and plasma membrane permeability [114]. Another investigation revealed that the essential oil compounds carvone, apiol and limonene (Figure 3) isolated from the seeds of the *Anathallis graveolens* (Pabst) F. Barros plant were synergistically active against *Aspergillus flavus*, which reduced ATPase and dehydrogenase synthesis, leading to fungal mitochondrial dysfunction and cell death induced by the accumulation ROS in *A. flavus* [115]. Moreover, Nakahara et al. [116] tested the combined activity of the EO compounds linalool and citronellal isolated from *C. nardus* against *Aspergillus* sp., *Eurotium* sp. and *Penicillium* sp., and found the combination to be synergistically fungicidal at a concentration of 112 mg/L [116]. Examples of fungicidal homosynergistic plant-derived EOs and their compounds are summarized in Table 1.

6.2. As Enhancers of Commercial Antifungal Agents

Plant-derived metabolites present in EOs are also capable of enhancing the antifungal action of existing synthetic chemical fungicidal agents. A study showed that the EO compound cinnamaldehyde potentiated the fungicidal action of fluconazole against *Aspergillus fumigatus* MTCC 2550 by reducing the MIC of the antifungal agent by up to 8-fold [117]. Gadban et al. [118] demonstrated that essential oil extracted from *Tagetesfilifolia* Lag. potentiated the fungicidal activity of difenoconazole, trifloxystrobin, cyproconazole and carbendazim up to 80% when used in combination against the phytopathogenic fungus *Colletotrichum truncatum* (Schweinitz) Andrus and W.D. Moore [118]. An investigation indicated that EO extracted from *Eupatorium adenophorum* leaves that consist of phytochemical compounds like 10Hα-9-oxo-agerophorone, 9-oxo-10, 11-dehydro-agerophorone and 10Hβ-9-oxo-agerophorone (Figure 3) was able to enhance the fungicidal action of mefenoxam and mancozebagainst *Pythium myriotylum* (Drechsler). According to the results of the study, the EO and mancozeb combination indicated the highest synergistic activity with a fungal mycelia growth rate of 100%, and light and transmission electron microscopic analysis revealed that the EO induced hyphae swelling, cell wall disruption, shortening of the cytoplasmic inclusion and degradation of plasma membrane and cytoplasmic organelles [119]. Camiletti et al. [120] tested and concluded the synergistic action of EO extracted from *Tagetes minuta* L., *Laurus nobilis* L. and *T. filifolia* with iprodione against a major crop-associated fungal pathogen known as *Sclerotium cepivorum* (Berkeley) Whetzel (withe rot). In the study, *T. minuta* in combination with iprodione showed the best synergistic activity, which induced 100% growth inhibition of the fungus. Furthermore, the study elucidated that phytochemical compounds anethole, phenylpropanoids, sphatulenol and estragole (Figure 3) were abundantly present in the EOs of the tested plants [120]. An investigation revealed that EO extracted from *Pogestemon patchouli* mediated partial synergism with synthetic antifungal agents like ketoconazole and amphotericin B against *A. niger* and *A. flavus* with a FICI ranging from 0.52 to 1 [121]. Examples of commercially available synthetic fungicidal agents used in combination with plant-derived essential oils and their compounds are summarized in Table 1.

7. Novel Developments in Synergistic Insecticidal and Fungicidal Plant-Derived Essential Oils

Recent developments and novel strategies have been implemented to enhance pesticidal and fungicidal actions of plant-based essential oils. A study indicated that the essential oil compound carvacrol (Figure 3) was able to synergistically interact with the crystalline proteins produced by *Bacillus thuringiensis* MPU B9 and MPU B54 strains to mediate larvicidal activity against *Cydia pomonella* (Linnaeus)(codling moth) and *S. exigua* (beet armyworm moth). The best synergistic larvicidal action was observed at a 1:25000 (MPU B54 protein to carvacrol) concentration ratio, which induced a 96.7% (±3.33%) mortality rate [122]. A similar study elucidated that EOs from *A. indica* containing azadirachtin and *Sinapis alba* were synergistically active against crop pests, like *Spodopteraexigua* (Hübner), *C. pomonella* and *Dendrolimus pini* (Linnaeus), when used in combination with bacterial crystalline toxins of *B. thuringiensis* MPU B9 isolate. Hence, the results of the study indicated a 2-fold increase in larvicidal activity of the combined agents [123]. An investigation conducted by Radha et al. [124] stated that essential oils extracted from *Chenopodium ambrosoides* and *Thymus vulgaris* induced synergism with fungal secretions released by *Beauveria bassiana*(Balsamo) Vuillemin to mediate insecticidal and repellent action against *Callosobruchus maculates* (Fabricius) (*Cowpea bruchid*). According to the results of the study, the highest synergistic interaction was observed with *Chenopodium* oil, which induced a 76% mortality rate of *C. maculates* larvae in 168 h after treatment [124]. Yang et al. [125] tested the insecticidal efficiency of polyethylene glycol-coated garlic essential oil against adult *T. castaneum* and found that these nanoparticles are capable of inducing 100% mortality [125]. An investigation demonstrated that essential oil purified from *Pelargonium graveolens* induced 40% mortality of the *Agrotis ipsilon* (Hufnagel) (dark sword-grass) moth when encapsulated and deployed with solid lipid nanoparticles [126]. Research conducted by Pierattini et al. [127] demonstrated that diatomaceous earth molecules worked synergistically to potentiate the insecticidal activity of *O. basilicum* and *Foeniculum vulgare* against *Sitophilus granaries* (Linnaeus). The combinative treatment indicated a synergistic co-toxicity coefficient that ranges from 1.36 to 3.35 for *F. vulgare* and *O. basilicum* [127]. A novel study demonstrated that orange essential oil interacted synergistically with a baculovirus known as the nucleopolyhedrosis virus to induce enhanced larvicidal activity against *S. littoralis* (the cotton leaf wormmoth) [128]. Furthermore, a novel study conducted by Al-alawi. [129] demonstrated that pine essential oil synergistically interacted with secretions of *B. bassiana* BAU016 fungal isolate to induce enhanced larvicidal activity against *Tetranychus urticae* (Koch) (two-spotted spider mite) [129].

An investigation conducted by Nasseri et al. [130] showed that the EO of *Zataria multiflora* mediated synergistic fungicidal action against *Aspergillus ochraceus*, *A. niger*, *A. flavus*, *Alternaria solani*, *Rhizoctonia solani* and *Rhizopus stolonifer* (Ehrenberg) when loaded and used with solid lipid nanoparticles. The study demonstrated that these combinations inhibited 54%–79% of fungal growth [130]. Luque-Alcaraz et al. [131] tested the antifungal efficiency of chitosan and *Schinus molle* (pepper tree) essential oil conjunctive bio-nanocomposites against *Aspergillus parasiticus* and observed a 40%–50% reduction in fungal cell viability [131]. A study indicated that *M. piperita* EO coated with gold nanoparticles induced synergistically enhanced antifungal activity against *A. flavus* [132]. An investigation revealed that *Satureja khuzestanica* (Jamza) essential oil encapsulated with chitosan nanoparticles induced enhanced fungicidal action against *R. stolonifer* [133]. A research study conducted by Kalagatur et al. [134] elucidated that chitosan nanoparticles mediated antifungal activity against the phytopathogenic fungus *Fusarium graminearum* (Schwabe) when incorporated with the EO of *Cymbopogon martini*, which indicated a MIC of 421.7 ± 27.14 and MFC of 618.3 ± 79.35 ppm. Scanning electron microscopic analysis in the study revealed detrimental changes in the fungal macroconidia and further elaborated antifungal action mechanisms like intracellular reactive oxygen species elevation, depletion of ergosterol content and lipid peroxidation. Moreover, the study revealed the abundance of geraniol (Figure 3) in the EO of *C. martini* [134]. Latha and Lal. [135] demonstrated that secretions produced by micro-algae were able to synergize and potentiate the antifungal action of thyme essential oil against the phytopathogenic fungus *Alternariabrassicae*, which causes a serious disease in pre-harvest and post-harvest broccoli crops [135]. A novel study showed that bioactive secretions of *Bacillus subtilis* B26 isolate synergistically enhanced the antifungal action of EOs obtained from myrtlewood, Leyland cypress needles, orange and lime when used in combination against phytopathogenicfungi *Ophiostoma perfectum*, *Trichoderma* spp. and *A. niger* [136]. Furthermore, a similar study elucidated that the essential oil extracted from *Zingiber officinale var. rubrum* induced enhanced fungicidal activity against an *A. niger* FNCC 6080 isolate when combined with the *Lactococcus lactis* produced bacteriocin lantibiotic known as nisin [137]. Examples of novel bioactive molecules used in combination with plant-derived essential oils and their compounds are summarized in Table 1.

Table 1. Plant-derived essential oils and their compounds combined with synergistic agents against paddy field insect pests and fungal pathogens.

Plant Source	EO Compound	Synergist Used with EO Compound	Insect Pest//Fungal Pathogen	Reference
N/S	Cinnamaldehyde	Eugenol	*L. sulphureus*	[111]
N/S	Cinnamaldehyde	Eugenol	*C. nardus*	[112]
Melaleuca alternifolia	Terpinolene, Terpinen-4-ol, δ-Terpinene,	α-pinene, 1,8-cineole, α-terpineol	*B. cinerea*	[114]
Anathallis graveolens	Carvone, Apiol	Limonene	*A. flavus*	[115]
Cymbopogon nardus	Linalool	Citronellal	*Aspergillus* sp., *Eurotium* sp., *Penicillium* sp.	[116]
Tagetes minuta, Laurus nobilis, Tagetes filifolia	Anethole, Phenylpropanoids, Sphatulenol, Estragole	Iprodione	*S. cepivorum*	[120]
N/S	Carvacrol	Crystalline proteins of *B. thuringiensis*	*C. pomonella, S. exigua*	[122]
Azadirachta indica *Rosmarinus officinalis*	Azadirachtin Camphor	Crystalline toxins of *B. thuringiensis* 1,8-cineole	*S. exigua, C. pomonella, D. pini* *T. ni*	[123] [90]
N/S	α-terpineol	Thymol	*C. partellus*	[92]
N/S	*Trans*-anethole, Citronellal	α-terpineol, and thymol	*S. litura*	[93]
N/S	Cinnamon oil	Rotenone	*S. litura*	[95]
N/S	Terpinolene	Safrole	*L. decemlineata, S. littoralis, R. padi, M. persicae*	[97]
Piper hispidinervum	γ–terpinene	Terpinen-4-ol	*A. fabae*	[96]
			S. littoralis	
Majorana hortensis	Cinnamaldehyde	Fluconazole	*A. fumigatus*	[117]

Table 1. *Cont.*

Plant Source	EO Compound	Synergist Used with EO Compound	Insect Pest//Fungal Pathogen	Reference
N/S N/S *Simmodsia chinesis*	10Hα-9-oxo-agerophorone, 9-oxo-10, 11-dehydro-agerophorone, 10Hβ-9-oxo-agerophorone Jojoba oil	Mefenoxam, Mancozeb Cyhalothrin	*P. myriotylum* *B. tabaci*	[119] [99]
Allium sativum	Garlic oil	Diuron		
Mentha piperita	Peppermint oil	Malathion		
Thymus vulgaris	N/S	Imidacloprid, Spirotetramat	*M. persicae*	[100]
Lavandula angustifolia	N/S	Deltamethrin	*S. frugiperda*	[102]
N/S	Linalool, Thymol	Decis® (25CE)	*S. Littoralis*	[103]
Ocimum basilicum	Linalool	Chlorpyrifos, Cypermethrin	*S. zeamais*	[104]
		Madex, Azatin, Helicovex	*R. dominica*	[105]
N/S	Garlic oil		*P. truncatus*	
N/S	Cannabidiol oil		*T. granarium*	

N/S: Not specified.

8. Concluding Remarks and Future Perspectives

The issue of synthetic pesticide, insecticide and fungicide resistance is expanding rapidly across the globe. Hence, the prospects for the application of existing pesticides and fungicides in the future have become challenging and uncertain. Plant-derived essential oils and their phytoconstituents are remarkable sources of novel bioactive compounds with broad-spectrum insecticidal and antifungal properties. These compounds can exert homosynergistic action or synergistically interact with other pest management agents or bioactive molecules. This review summarizes and interprets the findings of experimental work based on plant-based essential oils in combination with existing pesticidal, insecticidal and fungicidal agents, as well as novel bioactive natural and synthetic molecules, against insect pests and fungi responsible for the spoilage of crops. These essential oil combinations have shown remarkable results as agents with different mechanisms for overcoming pesticidal, insecticidal and fungicidal resistance. For instance, several studies have elucidated that these synergistic combinative compounds can significantly reduce the insect mortality rate and MIC/MFC of fungi. The efforts in synergy research have led to the discovery and production of novel pest management agents. However, the underlying modes of actions associated with synergistic essential oil products have not yet been fully exploited. Hence, the broadening of molecular and biochemical studies based on combined synergists of essential oils are needed to establish a better understanding and further exploitation of their toxicological responses and bioactivity in order to determine their true potency and safety in agricultural application. At present, the availability of experimental data based on essential oil synergists is limited and, therefore, further studies are needed in order to broaden and elucidate their novel action mechanisms in modifying pesticidal, insecticidal and fungicidal resistance. Moreover, studies are needed on insecticidal and fungicidal activities of fruit waste and botanical enzymes, like bromelain combinative synergists, with plant-derived essential oils.

Author Contributions: C.H.C. contributed by giving the original idea, concept. M.K.D. collaborated in designing and writing, data interpretation and drafting of the article. T.-J.K., A.F., A.S., C.M.C. and C.H.C. contributed to editing the final version of the draft to be published. All authors have read and agreed to the published version of the manuscript.

Funding: The authors would like to extend their sincere gratitude to Faculty of Science and Engineering, University of Nottingham, Malaysia for funding this research through pump priming grant F0013.54.04.

Institutional Review Board Statement: The study did not require any consent for publication.

Informed Consent Statement: The study did not require the consent of the participants.

Data Availability Statement: The following review was based on data extracted from published research articles available in all relevant databases with no limitation up to 10 June 2021.

Conflicts of Interest: The authors have no competing interests to declare.

Abbreviations

S. avermitilis: *Streptomyces avermitilis*; *S. spinose*: *Saccharopolyspora spinosa* (Mertz and Yao); spp.: Species (multiple); *P. variotii*: *Paecilomyces variotii*; AChE: Acetylcholinesterase; GABA: Gamma aminobutyric acid; CarEs: carboxy-lesterase; ATP: Adenosine triphosphate; GST: glutathione S-transferase; β: Beta; α: Alpha; δ: Delta; QS: Quorum sensing; EO: Essential oil; ROS: Reactive oxygen species; ECG: Electrocardiogram; LDH: Lactate dehydrogenase; *P. xylostella*: *Plutella xylostella* (Linnaeus); *C. partellus*: *Chilo partellus* (Swinhoe); *T. ni*: *Trichoplusia ni* (Hübner); *S. litura*: *Spodoptera litura* (F.); *S. oryzae*: *Sitophilus oryzae (L.)*; *B. rugimanus*: *Bruchus rugimanus* Bohem; *A. fabae*: *Aphis fabae* (Scopoli); *S. littoralis*: *Spodoptera littoralis* (Boisduval); *L. decemlineata*: *Leptinotarsa decemlineata* (Say); *R. padi*: *Rhopalosiphum padi*(Linnaeus); M. *persicae*: *Myzus persicae* (Sulzer); *T. castaneum*: *Tribolium castaneum* (Herbst); *B. tabaci*: *Bemisia tabaci* (Gennadius); *L. angustifolia*: *Lavandula angustifolia* (Miller); *S. frugiperda*; *Spodoptera frugiperda* (J.E. Smith); *S. zeamais*: *R. dominica*: *Rhyzopertha dominica* (Fabricius); *P. truncatus*: *Prostephanus truncates* (Horn); *T. granarium*: *Trogoderma granarium* (Everts); *F. oxysporum*: *Fusarium oxysporum* (Sacc.): *F. circinatum*: *Fusarium circinatum* (Nirenberg and O'Donnell); *C. gloeosporioides*: *Colletotrichum gloeosporioides* (Penz); *P. funiculosum*: *Penicillium funiculosum*; *M. racemosus*: *Mucor racemosus* (Fresenius); sp.: Species (single); *L. sulphureus*: *Laetiporussulphureus* (Bull.) Murrill; *S. pogostemonis*: *Synchytriumpogostemonis* S.D. Patil and Mahab; *A. niger*: *Aspergillus niger*; *A. alternata*; *Alternaria alternate* (Fries) Keissler; *C. gloeosporioides*: *Colletotrichumgloeosporioides* (Penzig); *L. theobromae*: *Lasiodiplodia theobromae* (Patouillard) Griffon and Maublanc; *P. viticola*: *Plasmopara viticola* (Berkeley and Curtis); *R. stolonifer*: *Rhizopus stolonifer* (Ehrenberg) Vuillemin; *P. viticola*: *Plasmopara viticola* (Berkeley and Curtis); *A. flavus*: *Aspergillus flavus*; *A. fumigatus*; *Aspergillus flavus*; *C. truncatum*: *Colletotrichum truncatum* (Schweinitz) Andrus and W.D. Moore; *P. myriotylum*: *Pythium myriotylum* (Drechsler); *S. cepivorum*: *Sclerotium cepivorum* (Berkeley) Whetzel; *L. lactis*: *Lactococcus lactis*; *T. minuta*: *Tagetes minuta* (Linnaeus); *C. pomonella*: *Cydia pomonella* (Linnaeus); *B. thuringiensis*: *Bacillus thuringiensis*; *S. exigua*: *Spodoptera exigua* (Hübner); *D. pini*: *Dendrolimus pini* (Linnaeus); *B. bassiana*: *Beauveria bassiana* (Balsamo) Vuillemin; *T. castaneum*: *Tribolium castaneum* (Herbst); *A. ipsilon*: *Agrotis ipsilon* (Hufnagel); *F. vulgare*: *Foeniculum vulgare* (Miller); *O. basilicum*: *Ocimum basilicum* (Linnaeus); *T. urticae*: *Tetranychus urticae* (Koch); *A. ochraceus*: *Aspergillus ochraceus*; *A. brassicae*: *Alternaria brassicae*; *A. solani*: *Alternaria solani*; *A. ochraceus: Aspergillus ochraceus*; *R. solani*: *Rhizoctonia solani*; *R. stolonifer*: *Rhizopus stolonifer*(Ehrenberg) Vuillemin; *A. parasiticus*: *Aspergillus parasiticus*; *C. maculates*: *Callosobruchus maculates* (Fabricius) µg: Microgram; >: Greater than; mg/L: Milligram per liter; µm: Micrometer; µL: Microliter; ppm: Part per million; µL/L: Microliter per liter; FICI: Fractional inhibitory concentration index; <: Less than; mL: Milliliter; µL/mL: Microliter per milliliter; USD: United States Dollars; USA: United States of America; MIC: Minimum inhibitory concentration; MTCC: Microbial-Type Culture Collection and Gene Bank; MFC: Minimum fungicidal concentration; FNCC: Food and Nutrition Culture Collection.

References

1. Van Dijk, M.; Morley, T.; Rau, M.; Saghai, Y. A meta-analysis of projected global food demand and population at risk of hunger for the period 2010–2050. *Nat. Food* **2021**, *2*, 494–501. [CrossRef]
2. Popp, J.; Pető, K.; Nagy, J. Pesticide productivity and food security. A review. *Agron. Sustain. Dev.* **2012**, *33*, 243–255. [CrossRef]

3. Zhang, W. Global pesticide use: Profile, trend, cost/benefit and more. *Proc. Int. Acad. Ecol. Environ. Sci.* **2018**, *8*, 1–27.
4. Sharma, A.; Kumar, V.; Shahzad, B.; Tanveer, M.; Sidhu, G.; Handa, N.; Kohli, S.; Yadav, P.; Bali, A.; Parihar, R.; et al. Worldwide pesticide usage and its impacts on ecosystem. *SN Appl. Sci.* **2019**, *1*, 1446. [CrossRef]
5. Aktar, W.; Sengupta, D.; Chowdhury, A. Impact of pesticides use in agriculture: Their benefits and hazards. *Interdiscip. Toxicol.* **2009**, *2*, 1–12. [CrossRef]
6. Saiyed, H.N.; Sadhu, H.G.; Bhatnagar, V.K.; Dewan, A.; Venkaiah, K.; Kashyap, S.K. Cardiac toxicity following short-term exposure to methomyl in spraymen and rabbits. *Hum. Exp. Toxicol.* **1992**, *11*, 93–97. [CrossRef]
7. Sanders, H.O. *Toxicity of Pesticides to the Crustacean Gammarus Lacustris*; Technical Papers of the Bureau of Sport Fisheries and Wildlife 1969, No. 25; US Department of Interior Fish and Wildlife Service: Washington, DC, USA, 1969.
8. United Nations Sustainable Development. Goal 2: Zero Hunger. Available online: https://www.un.org/sustainabledevelopment/hunger/ (accessed on 10 June 2021).
9. Lourenço, S.; Moldão-Martins, M.; Alves, V. Antioxidants of natural plant origins: From sources to food industry applications. *Molecules* **2019**, *24*, 4132. [CrossRef]
10. Dhifi, W.; Bellili, S.; Jazi, S.; Bahloul, N.; Mnif, W. Essential oils' chemical characterization and investigation of some biological activities: A critical review. *Medicines* **2016**, *3*, 25. [CrossRef]
11. Bhavaniramya, S.; Vishnupriya, S.; Al-Aboody, M.; Vijayakumar, R.; Baskaran, D. Role of essential oils in food safety: Antimicrobial and antioxidant applications. *Grain Oil Sci. Technol.* **2019**, *2*, 49–55. [CrossRef]
12. Pott, D.; Osorio, S.; Vallarino, J. From central to specialized metabolism: An overview of some secondary compounds derived from the primary metabolism for their role in conferring nutritional and organoleptic characteristics to fruit. *Front. Plant. Sci.* **2019**, *10*, 835. [CrossRef]
13. Preedy, V. *Essential Oils in Food Preservation, Flavor and Safety 2016*; Academic Press: Cambridge, MA, USA, 2015.
14. Tripathi, A.K.; Upadhyay, S.; Bhuiyan, M.; Bhattacharya, P.R. A review on prospects of essential oils as biopesticide in insect-pest management. *J. Pharmacogn. Phytother.* **2009**, *1*, 52–63.
15. Sharmeen, J.B.; Mahomoodally, F.M.; Zengin, G.; Maggi, F. Essential oils as natural sources of fragrance compounds for cosmetics and cosmeceuticals. *Molecules* **2021**, *26*, 666. [CrossRef] [PubMed]
16. Svoboda, K. Investigation of volatile oil glands of *Satureja hortensis* L. (summer savory) and phytochemical comparison of different varieties. *Int. J. Aromather.* **2003**, *13*, 196–202. [CrossRef]
17. Fahn, A. Structure and function of secretory cells. *Adv. Bot. Res.* **2000**, *31*, 37–75.
18. Asgar, E.; Jalal, J.S. A review on recent research results on bio-effects of plant essential oils against major coleopteran insect pests. *Toxin Rev.* **2015**, *34*, 76–91.
19. WHO. *International Code of Conduct on the Distribution and Use of Pesticides: Guidelines for the Registration of Pesticides*; World Health Organization: Rome, Italy, 2010.
20. FAO. *Manual on the Submission and Evaluation of Pesticide Residues Data for the Estimation of Maximum Residue Levels in Food and Feed*; Food and Agriculture Organization: Rome, Italy, 2002.
21. Damalas, C.A.; Eleftherohorinos, I.G. Pesticide exposure, safety issues, and risk assessment indicators. *Int. J. Environ. Res. Public Health* **2011**, *8*, 1402–1419. [CrossRef]
22. Wińska, K.; Mączka, W.; Łyczko, J.; Grabarczyk, M.; Czubaszek, A.; Szumny, A. Essential oils as antimicrobial Agents—Myth Or real alternative? *Molecules* **2019**, *24*, 2130. [CrossRef]
23. Yan, Y.; Liu, Q.; Jacobsen, S.; Tang, Y. The impact and prospect of natural product discovery in agriculture. *EMBO Rep.* **2018**, *19*, e46824. [CrossRef]
24. Boulogne, I.; Petit, P.; Ozier-Lafontaine, H.; Desfontaines, L.; Loranger-Merciris, G. Insecticidal and antifungal chemicals produced by plants: A review. *Environ. Chem. Lett.* **2012**, *10*, 325–347. [CrossRef]
25. Bynum, B. Shedding new light on the story of penicillin. *Lancet* **2007**, *369*, 1991–1992. [CrossRef]
26. Sparks, T.; Hahn, D.; Garizi, N. Natural products, their derivatives, mimics and synthetic equivalents: Role in agrochemical discovery. *Pest Manag. Sci.* **2016**, *73*, 700–715. [CrossRef]
27. Rao, G.V.; Rupela, O.P.; Rao, V.R.; Reddy, Y.V. Role of biopesticides in crop protection: Present status and future prospects. *Indian J. Plant. Prot.* **2007**, *35*, 1–9.
28. Miller, G. *Living in the Environment*, 12th ed.; Thomson Learning: Belmont, CA, USA, 2002.
29. Maheshwari, R. *Fungi: Experimental Methods in Biology*, 2nd ed.; CRC Press: Boca Raton, FL, USA, 2012.
30. Pandey, A.K.; Sain, S.K.; Singh, P. A perspective on integrated disease management in agriculture. *Bio Bull.* **2016**, *2*, 13–29.
31. Kirst, H. The spinosyn family of insecticides: Realizing the potential of natural products research. *J. Antibiot.* **2010**, *63*, 101–111. [CrossRef] [PubMed]
32. Áy, Z.; Mihály, R.; Cserháti, M.; Kótai, É.; Pauk, J. The effect of high concentrations of glufosinate ammonium on the yield components of transgenic spring wheat (*Triticum aestivum* L.) constitutively expressing the bar gene. *Sci. World J.* **2012**, *2012*, 657945. [CrossRef]
33. Hoerlein, G. Glufosinate (phosphinothricin), a natural amino acid with unexpected herbicidal properties. *Rev. Environ. Contam. Toxicol.* **1994**, *138*, 73–145. [CrossRef]

34. Owen, W.; Meyer, K.; Slanec, T.; Meyer, S.; Wang, N.; Fitzpatrick, G.; Niyaz, N.; Nugent, J.; Ricks, M.; Rogers, R.; et al. Synthesis and biological activity of analogs of the antifungal antibiotic uk-2a. iii. impact of modifications to the macrocycle isobutyryl ester position. *Pest Manag. Sci.* **2019**, *76*, 277–286. [CrossRef] [PubMed]
35. Tongnuanchan, P.; Benjakul, S. Essential oils: Extraction, bioactivities, and their uses for food preservation. *J. Food Sci.* **2014**, *79*, 1231–1249. [CrossRef]
36. Masango, P. Cleaner production of essential oils by steam distillation. *J. Clean. Prod.* **2005**, *13*, 833–839. [CrossRef]
37. Bakkali, F.; Averbeck, S.; Averbeck, D.; Idaomar, M. Biological effects of essential oils—A review. *Food Chem. Toxicol.* **2008**, *46*, 446–475. [CrossRef]
38. Mohamed, A.A.; El-Emary, G.A.; Ali, H.F. Influence of some citrus essential oils on cell viability, glutathione-s-transferase and lipid peroxidation in ehrlich ascites carcinoma cells. *J. Am. Sci.* **2010**, *6*, 820–826.
39. Ngoh, S.; Choo, L.; Pang, F.; Huang, Y.; Kini, M.; Ho, S. Insecticidal and repellent properties of nine volatile constituents of essential oils against the american cockroach, *Periplaneta americana* (L.). *Pestic. Sci.* **1998**, *54*, 261–268. [CrossRef]
40. Butnariu, M.; Sarac, I. Essential oils from plants. *J. Biotechnol. Biomed. Sci.* **2018**, *1*, 35–43. [CrossRef]
41. Borzoui, E.; Naseri, B.; Abedi, Z.; Karimi-Pormehr, M. Lethal and sublethal effects of essential oils from *Artemisia khorassanica* and *Vitex pseudo*-negundo against *Plodia interpunctella* (Lepidoptera: Pyralidae). *Environ. Entomol.* **2016**, *45*, 1220–1226. [CrossRef]
42. Daferera, D.; Ziogas, B.; Polissiou, M. GC-MS analysis of essential oils from some greek aromatic plants and their fungitoxicity on *Penicillium digitatum*. *J. Agric. Food Chem.* **2000**, *48*, 2576–2581. [CrossRef] [PubMed]
43. Hermans, P.E.; Keys, T.F. Antifungal agents used for deep-seated mycotic infections. *Mayo Clin. Proc.* **1983**, *58*, 223–231. [PubMed]
44. Kyle, A.; Dahl, M. Topical therapy for fungal infections. *Am. J. Clin. Dermatol.* **2004**, *5*, 443–451. [CrossRef] [PubMed]
45. Niroumand, M.C.; Farzaei, M.; Karimpour Razkenari, E.; Amin, G.; Khanavi, M.; Akbarzadeh, T.; Shams-Ardekani, M. An evidence-based review on medicinal plants used as insecticide and insect repellent in traditional iranian medicine. *Iran. Red Crescent Med. J.* **2016**, *18*, e22361.
46. Kim, J.; Jung, C.; Koh, Y.; Lee, S. Molecular, biochemical and histochemical characterization of two acetylcholinesterase cDNAs from the German cockroach *Blattella germanica*. *Insect Mol. Biol.* **2006**, *15*, 513–522. [CrossRef]
47. Pezzementi, L.; Rowland, M.; Wolfe, M.; Tsigelny, I. Inactivation of an invertebrate acetylcholinesterase by sulfhydryl reagents: The roles of two cysteines in the catalytic gorge of the enzyme. *Invertebr. Neurosci.* **2006**, *6*, 47–55. [CrossRef]
48. Pang, Y.; Singh, S.; Gao, Y.; Lassiter, T.; Mishra, R.; Zhu, K.; Brimijoin, S. Selective and irreversible inhibitors of aphid acetylcholinesterases: Steps toward human-safe insecticides. *PLoS ONE* **2009**, *4*, e4349. [CrossRef]
49. Polsinelli, G.; Singh, S.; Mishra, R.; Suranyi, R.; Ragsdale, D.; Pang, Y.; Brimijoin, S. Insect-specific irreversible inhibitors of acetylcholinesterase in pests including the bed bug, the eastern yellowjacket, German and American cockroaches, and the confused flour beetle. *Chem. Biol. Interact.* **2010**, *187*, 142–147. [CrossRef]
50. Pang, Y.; Brimijoin, S.; Ragsdale, D.W.; Yan Zhu, K.; Suranyi, R. Novel and viable acetylcholinesterase target site for developing effective and environmentally safe insecticides. *Curr. Drug Targets* **2012**, *13*, 471–482. [CrossRef]
51. Gnagey, A.; Forte, M.; Rosenberry, T. Isolation and characterization of acetylcholinesterase from Drosophila. *J. Biol. Chem.* **1987**, *262*, 13290–13298. [CrossRef]
52. Bourguet, D.; Roig, A.; Toutant, J.; Arpagaus, M. Analysis of molecular forms and pharmacological properties of acetylcholinesterase in several mosquito species. *Neurochem. Int.* **1997**, *31*, 65–72. [CrossRef]
53. Marcel, V.; Palacios, L.; Pertuy, C.; Masson, P.; Fournier, D. Two invertebrate acetylcholinesterases show activation followed by inhibition with substrate concentration. *Biochem. J.* **1998**, *329*, 329–334. [CrossRef]
54. Marrs, T.; Maynard, R. Neurotranmission systems as targets for toxicants: A review. *Cell Biol. Toxicol.* **2013**, *29*, 381–396. [CrossRef] [PubMed]
55. Re, L.; Barocci, S.; Sonnino, S.; Mencarelli, A.; Vivani, C.; Paolucci, G.; Scarpantonio, A.; Rinaldi, L.; Mosca, E. Linalool modifies the nicotinic receptor-ion channel kinetics at the mouse neuromuscular junction. *Pharmacol. Res.* **2000**, *42*, 177–181. [CrossRef] [PubMed]
56. Kostyukovsky, M.; Rafaeli, A.; Gileadi, C.; Demchenko, N.; Shaaya, E. Activation of octopaminergic receptors by essential oil constituents isolated from aromatic plants: Possible mode of action against insect pests. *Pest Manag. Sci.* **2002**, *58*, 1101–1106. [CrossRef]
57. Evans, P.D. Biogenic amines in the insect nervous system. *Adv. Insect Physiol.* **1980**, *15*, 317–473.
58. Evans, P.D. Multiple receptor types for octopamine in the locust. *J. Physiol.* **1981**, *318*, 99–122. [CrossRef]
59. Farooqui, T. Octopamine-mediated neuromodulation of insect senses. *Neurochem. Res.* **2007**, *32*, 1511–1529. [CrossRef]
60. Ohta, H.; Ozoe, Y. Molecular signalling, pharmacology, and physiology of octopamine and tyramine receptors as potential insect pest control targets. *Adv. Insect Physiol.* **2014**, *46*, 73–166.
61. Lange, A.; Orchard, I. Identified octopaminergic neurons modulate contractions of locust visceral muscle via adenosine 3′,5′-monophosphate (cyclic AMP). *Brain Res.* **1986**, *363*, 340–349. [CrossRef]
62. Nathanson, J. Octopamine receptors, adenosine 3′,5′-monophosphate, and neural control of firefly flashing. *Science* **1979**, *203*, 65–68. [CrossRef] [PubMed]
63. Orchard, I.; Carlisle, J.; Loughton, B.; Gole, J.; Downer, R. In vitro studies on the effects of octopamine on locust fat body. *Gen. Comp. Endocrinol.* **1982**, *48*, 7–13. [CrossRef]

64. Ben-Ari, Y.; Khalilov, I.; Kahle, K.; Cherubini, E. The GABA excitatory/inhibitory shift in brain maturation and neurological disorders. *Neuroscientist* **2012**, *18*, 467–486. [CrossRef]
65. Sattelle, D. GABA receptors of insects. *Adv. Insect Physiol.* **1990**, *22*, 1–113.
66. Sigel, E.; Steinmann, M. Structure, function, and modulation of GABAA receptors. *J. Biol. Chem.* **2012**, *287*, 40224–40231. [CrossRef]
67. Yu, H.; Xu, J.; Wang, X.; Ma, Y.; Yu, D.; Fei, D.; Zhang, S.; Wang, W. Identification of four ATP-binding cassette transporter genes in *Cnaphalocrocis medinalis* and their expression in response to insecticide treatment. *J. Insect Sci.* **2017**, *17*, 44. [CrossRef]
68. Merzendorfer, H. Chitin synthesis inhibitors: Old molecules and new developments. *Insect Sci.* **2012**, *20*, 121–138. [CrossRef]
69. Campolo, O.; Giunti, G.; Russo, A.; Palmeri, V.; Zappalà, L. Essential oils in stored product insect pest control. *J. Food Qual.* **2018**, *2018*, 6906105. [CrossRef]
70. Hyldgaard, M.; Mygind, T.; Meyer, R. Essential oils in food preservation: Mode of action, synergies, and interactions with food matrix components. *Front. Microbiol.* **2012**, *3*, 12. [CrossRef]
71. Hu, Y.; Zhang, J.; Kong, W.; Zhao, G.; Yang, M. Mechanisms of antifungal and anti-aflatoxigenic properties of essential oil derived from turmeric (*Curcuma longa* L.) on aspergillus flavus. *Food Chem.* **2017**, *220*, 1–8. [CrossRef]
72. Kalemba, D.; Kunicka, A. Antibacterial and antifungal properties of essential oils. *Curr. Med. Chem.* **2003**, *10*, 813–829. [CrossRef] [PubMed]
73. Prakash, B.; Singh, P.; Kedia, A.; Dubey, N. Assessment of some essential oils as food preservatives based on antifungal, antiaflatoxin, antioxidant activities and in vivo efficacy in food system. *Food Res. Int.* **2012**, *49*, 201–208. [CrossRef]
74. Lang, G.; Buchbauer, G. A review on recent research results (2008–2010) on essential oils as antimicrobials and antifungals. A review. *Flavour Fragr. J.* **2011**, *27*, 13–39. [CrossRef]
75. Wu, X.Z.; Cheng, A.X.; Sun, L.M.; Lou, H.X. Effect of plagiochin E, an antifungal macrocyclic bis (bibenzyl), on cell wall chitin synthesis in *Candida albicans*. *Acta Pharmacol. Sin.* **2008**, *12*, 78–1485. [CrossRef] [PubMed]
76. Kerekes, E.; Deák, É.; Takó, M.; Tserennadmid, R.; Petkovits, T.; Vágvölgyi, C.; Krisch, J. Anti-biofilm forming and anti-quorum sensing activity of selected essential oils and their main components on food-related micro-organisms. *J. Appl. Microbiol.* **2013**, *115*, 933–942. [CrossRef] [PubMed]
77. Rajput, S.B.; Karuppayil, S.M. Small molecules inhibit growth, viability and ergosterol biosynthesis in *Candida albicans*. *Springerplus* **2013**, *2*, 26. [CrossRef] [PubMed]
78. Ahmad, A.; Khan, A.; Kumar, P.; Bhatt, R.P.; Manzoor, N. Antifungal activity of *Coriaria nepalensis* essential oil by disrupting ergosterol biosynthesis and membrane integrity against *Candida*. *Yeast* **2011**, *28*, 611–617. [CrossRef]
79. Freires, I.; Murata, R.; Furletti, V.; Sartoratto, A.; Alencar, S.; Figueira, G.; de Oliveira Rodrigues, J.; Duarte, M.; Rosalen, P. *Coriandrum Sativum* L. (coriander) essential oil: Antifungal activity and mode of action on *Candida* spp., and molecular targets affected in human whole-genome expression. *PLoS ONE* **2014**, *9*, e99086. [CrossRef] [PubMed]
80. Nazzaro, F.; Fratianni, F.; Coppola, R.; Feo, V. Essential oils and antifungal activity. *Pharmaceuticals* **2017**, *10*, 86. [CrossRef] [PubMed]
81. Lee, H.; Ahn, Y. Growth-inhibiting effects of cinnamomum cassia bark-derived materials on human intestinal bacteria. *J. Agric. Food Chem.* **1998**, *46*, 8–12. [CrossRef]
82. Bang, K.; Lee, D.; Park, H.; Rhee, Y. Inhibition of fungal cell wall synthesizing enzymes bytrans-cinnamaldehyde. *Biosci. Biotechnol. Biochem.* **2000**, *64*, 1061–1063. [CrossRef] [PubMed]
83. Williams, P. Quorum sensing, communication and cross-kingdom signalling in the bacterial world. *Microbiology* **2007**, *153*, 3923–3938. [CrossRef] [PubMed]
84. Winans, S.C.; Bassler, B.L. Mob psychology. *J. Bacteriol.* **2002**, *184*, 873–883. [CrossRef]
85. Li, Z.; Nair, S. Quorum sensing: How bacteria can coordinate activity and synchronize their response to external signals? *Protein Sci.* **2012**, *21*, 1403–1417. [CrossRef]
86. Lumjiaktase, P.; Aguilar, C.; Battin, T.; Riedel, K.; Eberl, L. Construction of self-transmissible green fluorescent protein-based biosensor plasmids and their use for identification of N-acyl homoserine-producing bacteria in lake sediments. *Appl. Environ. Microbiol.* **2010**, *76*, 6119–6127. [CrossRef]
87. Uroz, S.; D'Angelo-Picard, C.; Carlier, A.; Elasri, M.; Sicot, C.; Petit, A.; Oger, P.; Faure, D.; Dessaux, Y. Novel bacteria degrading N-acylhomoserine lactones and their use as quenchers of quorum-sensing-regulated functions of plant-pathogenic bacteria. *Microbiology* **2003**, *149*, 1981–1989. [CrossRef]
88. Rasmussen, T.; Skindersoe, M.; Bjarnsholt, T.; Phipps, R.; Christensen, K.; Jensen, P.; Andersen, J.; Koch, B.; Larsen, T.; Hentzer, M.; et al. Identity and effects of quorum-sensing inhibitors produced by penicillium species. *Microbiology* **2005**, *151*, 1325–1340. [CrossRef]
89. Kumrungsee, N.; Pluempanupat, W.; Koul, O.; Bullangpoti, V. Toxicity of essential oil compounds against diamondback moth, plutella xylostella, and their impact on detoxification enzyme activities. *J. Pest Sci.* **2014**, *87*, 721–729. [CrossRef]
90. Tak, J.; Isman, M. Enhanced cuticular penetration as the mechanism for synergy of insecticidal constituents of rosemary essential oil in *Trichoplusia ni*. *Sci. Rep.* **2015**, *5*, 1–10. [CrossRef] [PubMed]
91. Tak, J.; Isman, M. Penetration-enhancement underlies synergy of plant essential oil terpenoids as insecticides in the cabbage looper, *Trichoplusia ni*. *Sci. Rep.* **2017**, *7*, 1–11. [CrossRef]

92. Singh, R.; Koul, O.; Rup, P.; Jindal, J. Toxicity of some essential oil constituents and their binary mixtures against *Chilo partellus* (lepidoptera: Pyralidae). *Int. J. Trop. Insect Sci.* **2009**, *29*, 93–101. [CrossRef]
93. Hummelbrunner, L.; Isman, M. Acute, sublethal, antifeedant, and synergistic effects of monoterpenoid essential oil compounds on the tobacco cutworm, *Spodoptera litura* (Lep., Noctuidae). *J. Agric. Food Chem.* **2001**, *49*, 715–720. [CrossRef]
94. Liu, C.; Mishra, A.; Tan, R.; Tang, C.; Yang, H.; Shen, Y. Repellent and insecticidal activities of essential oils from *Artemisia princeps* and *Cinnamomum camphora* and their effect on seed germination of wheat and broad bean. *Bioresour. Technol.* **2006**, *97*, 1969–1973. [CrossRef] [PubMed]
95. Li, Z.; Huang, R.; Li, W.; Cheng, D.; Mao, R.; Zhang, Z. Addition of cinnamon oil improves toxicity of rotenone tospodoptera litura (lepidoptera: Noctuidae) larvae. *Fla. Entomol.* **2017**, *100*, 515–521. [CrossRef]
96. Abbassy, M.; Abdelgaleil, S.; Rabie, R. Insecticidal and synergistic effects of Majorana hortensis essential oil and some of its major constituents. *Entomol. Exp. Appl.* **2009**, *131*, 225–232. [CrossRef]
97. Andrés, M.; Rossa, G.; Cassel, E.; Vargas, R.; Santana, O.; Díaz, C.; González-Coloma, A. Biocidal effects of piper hispidinervum (piperaceae) essential oil and synergism among its main components. *Food Chem. Toxicol.* **2017**, *109*, 1086–1092. [CrossRef] [PubMed]
98. Liang, J.; Xu, J.; Yang, Y.; Shao, Y.; Zhou, F.; Wang, J. Toxicity and synergistic effect of elsholtzia ciliata essential oil and its main components against the adult and larval stages of tribolium castaneum. *Foods* **2020**, *9*, 345. [CrossRef]
99. El-Meniawi, F.A.; Ismail, S.M. Toxic and biochemical impact of certain plant essential oils on *Bemisia tabaci* genn. (Hom., Aleyrodidae). *J. Pest Cont. Environ. Sci.* **2006**, *14*, 18–99.
100. Faraone, N.; Hillier, N.; Cutler, G. Plant essential oils synergize and antagonize toxicity of different conventional insecticides against *Myzus persicae* (hemiptera: Aphididae). *PLoS ONE* **2015**, *10*, e0127774. [CrossRef] [PubMed]
101. Martín-López, B.; Varela, I.; Marnotes, S.; Cabaleiro, C. Use of oils combined with low doses of insecticide for the control ofMyzus persicae and PVY epidemics. *Pest Manag. Sci.* **2006**, *62*, 372–378. [CrossRef] [PubMed]
102. Silva, S.; Cunha, J.; Carvalho, S.; Zandonadi, C.; Martins, R.; Chang, R. Ocimum basilicum essential oil combined with deltamethrin to improve the management of *Spodoptera frugiperda*. *Ciência Agrotecnologia* **2017**, *41*, 665–675. [CrossRef]
103. Macedo Silva, S.; Cunha, J.; Zandonadi, C.; Assunção, H.; Gregorio Marques, M. Synergistic effects of binary mixtures of linalool with pyrethroids against fall armyworm. *Biosci. J.* **2020**, *36*, 228–237. [CrossRef]
104. Ismail, S.M. Synergistic efficacy of plant essential oils with cypermethrin and chlorpyrifos against *Spodoptera littoralis*, field populations in egypt. *Int. J. Adv. Biol. Biomed. Res.* **2021**, *9*, 128–137.
105. Mantzoukas, S.; Kalyvas, N.; Ntoukas, A.; Lagogiannis, I.; Farsalinos, K.; Eliopoulos, P.; Poulas, K. Combined toxicity of cannabidiol oil with three bio-pesticides against adults of *Sitophilus zeamais, Rhyzopertha dominica, Prostephanus truncatus* and *Trogoderma granarium*. *Int. J. Environ. Res. Public Health* **2020**, *17*, 6664. [CrossRef] [PubMed]
106. Perczak, A.; Gwiazdowska, D.; Marchwińska, K.; Juś, K.; Gwiazdowski, R.; Waśkiewicz, A. Antifungal activity of selected essential oils against fusarium culmorum and f. graminearum and their secondary metabolites in wheat seeds. *Arch. Microbiol.* **2019**, *201*, 1085–1097. [CrossRef] [PubMed]
107. Figueroa, M.; López, D.; Anaya, L.; López, D.; Quijano, R.; Ovando, A. Chitosan composite films: Physicochemical characterization and their use as coating in papaya Maradol stored at room temperature. *Emir. J. Food Agric.* **2017**, *29*, 779–791. [CrossRef]
108. Nardoni, S.; D'Ascenzi, C.; Caracciolo, I.; Mannaioni, G.; Papini, R.; Pistelli, L.; Najar, B.; Mancianti, F. Activity of selected essential oils on spoiling fungi cultured from marzolino cheese. *Ann. Agric. Environ. Med.* **2018**, *25*, 280–284. [CrossRef] [PubMed]
109. Bedoya-Serna, C.; Dacanal, G.; Fernandes, A.; Pinho, S. Antifungal activity of nanoemulsions encapsulating oregano (*Origanum vulgare*) essential oil: In vitro study and application in Minas Padrão cheese. *Braz. J. Microbiol.* **2018**, *49*, 929–935. [CrossRef] [PubMed]
110. Bounar, R.; Krimat, S.; Boureghda, H.; Dob, T. Chemical analyses, antioxidant and antifungal effects of oregano and thyme essential oils alone or in combination against selected Fusarium species. *Int. Food Res. J.* **2020**, *27*, 66–77.
111. Yen, T.; Chang, S. Synergistic effects of cinnamaldehyde in combination with eugenol against wood decay fungi. *Bioresour. Technol.* **2008**, *99*, 232–236. [CrossRef]
112. Hartati, S.; Sukamto, N.; Karyani, N.; Zulhisnain, N. Efficacy of single formula of clove, eucalyptus, neem and citronella oil against budok disease of patchouli. *Bul. Penelit. Tanam. Rempah Dan Obat* **2018**, *28*, 153. [CrossRef]
113. Sukatta, U.; Haruthaithanasa, V.; Chantarapanont, W.; Dilokkunanant, U.; Suppakul, P. Antifungal activity of clove and cinnamon oil and their synergistic against postharvest decay fungi of grape in vitro. *Kasetsart J. Nat. Sci.* **2008**, *42*, 169–174.
114. Yu, D.; Wang, J.; Shao, X.; Xu, F.; Wang, H. Antifungal modes of action of tea tree oil and its two characteristic components against *Botrytis cinerea*. *J. Appl. Microbiol.* **2015**, *119*, 1253–1262. [CrossRef] [PubMed]
115. Tian, J.; Ban, X.; Zeng, H.; He, J.; Chen, Y.; Wang, Y. The mechanism of antifungal action of essential oil from dill (*Anethum graveolens* L.) on aspergillus flavus. *PLoS ONE* **2012**, *7*, e30147.
116. Nakahara, K.; Alzoreky, N.; Yoshihashi, T.; Nguyen, H.; Trakoontivakorn, G. Chemical composition and antifungal activity of essential oil from *Cymbopogon nardus* (citronella grass). *Jpn. Agric. Res. Q. JARQ* **2013**, *37*, 249–252. [CrossRef]
117. Khan, M.; Ahmad, I. Antifungal activity of essential oils and their synergy with fluconazole against drug-resistant strains of *Aspergillus fumigatus* and *Trichophyton rubrum*. *Appl. Microbiol. Biotechnol.* **2011**, *90*, 1083–1094. [CrossRef] [PubMed]

118. Gadban, L.C.; Camiletti, B.X.; Bigatton, E.D.; Distéfano, S.G.; Lucini, E.I. Combinations of *Tagetes filifolia* lag. essential oil with chemical fungicides to control *Colletotrichum truncatum* and their effects on the biocontrol agent *Trichoderma harzianum*. *J. Plant Prot. Res.* **2020**, *60*, 41–50.
119. Liu, X.; Yan, D.; Ouyang, C.; Yang, D.; Wang, Q.; Li, Y.; Guo, M.; Cao, A. Oils extracted from *Eupatorium adenophorum* leaves show potential to control *Phythium myriotylum* in commercially-grown ginger. *PLoS ONE* **2017**, *12*, e0176126.
120. Camiletti, B.; Asensio, C.; Gadban, L.; Pecci, M.; Conles, M.; Lucini, E. Essential oils and their combinations with iprodione fungicide as potential antifungal agents against withe rot (*Sclerotium cepivorum* berk) in garlic (*Allium sativum* L.) crops. *Ind. Crop. Prod.* **2016**, *85*, 117–124. [CrossRef]
121. Shin, S. Anti-aspergillus activities of plant essential oils and their combination effects with ketoconazole or amphotericin b. *Arch. Pharmacal Res.* **2003**, *26*, 389–393. [CrossRef]
122. Konecka, E.; Kaznowski, A.; Grzesiek, W.; Nowicki, P.; Czarniewska, E.; Baranek, J. Synergistic interaction between carvacrol and *Bacillus thuringiensis* crystalline proteins against *Cydia pomonella* and *Spodoptera exigua*. *BioControl* **2020**, *65*, 447–460. [CrossRef]
123. Konecka, E.; Kaznowski, A.; Tomkowiak, D. Insecticidal activity of mixtures of *Bacillus thuringiensis* crystals with plant oils of *Sinapis alba* and *Azadirachta indica*. *Ann. Appl. Biol.* **2019**, *174*, 364–371. [CrossRef]
124. Radha, R.; Murugan, K.; Wei, H.; Amerasan, D.; Madhiyazhagan, P.; Chen, F.; Kovendan, K.; Nataraj, T.; Nareshkumar, A.; Hwang, J.; et al. Insecticidal activity of essential oils and entomopathogenic fungiagainst cowpea bruchid, *Callosobruchus maculatus* (f.) (insecta:coleoptera: Bruchidae). *Int. J. Curr. Innov. Res.* **2014**, *1*, 12–19.
125. Yang, F.L.; Li, X.G.; Zhu, F.; Lei, C.L. Structural characterization of nanoparticles loaded with garlic essential oil and their insecticidal activity against *Tribolium castaneum* (Herbst) (coleoptera: Tenebrionidae). *J. Agric. Food Chem.* **2009**, *57*, 10156–10162. [CrossRef] [PubMed]
126. Adel, M.M.; Salem, N.Y.; Abdel-Aziz, N.F.; Ibrahim, S.S. Application of new nano pesticide Geranium oil loadedsolid lipid nanoparticles for control the black cutworm *Agrotis ipsilon* (Hub.) (Lepi., Noctuidae). *EurAsian J. BioSci.* **2019**, *13*, 1453–1461.
127. Pierattini, E.; Bedini, S.; Venturi, F.; Ascrizzi, R.; Flamini, G.; Bocchino, R.; Girardi, J.; Giannotti, P.; Ferroni, G.; Conti, B. Sensory quality of essential oils and their synergistic effect with diatomaceous earth, for the control of stored grain insects. *Insects* **2019**, *10*, 114. [CrossRef] [PubMed]
128. Sayed, W.; El-Bendary, H.; El-Helaly, A. Increasing the efficacy of the cotton leaf worm *Spodoptera littoralis* nucleopolyhedrosis virus using certain essential oils. *Egypt. J. Biol. Pest Control* **2020**, *30*, 8. [CrossRef]
129. Al-alawi, M.S. Evaluation of Jordanian isolates of *Beauveria bassiana* (Balsamo) Vuillemin and their interaction with essential plant oils when combined for the two spotted spider mite, *Tetranychus urticae* koch control. *Adv. Environ. Biol.* **2019**, *13*, 4–10.
130. Nasseri, M.; Golmohammadzadeh, S.; Arouiee, H.; Jaafari, M.R.; Neamati, H. Antifungal activity of *Zataria multiflora* essential oil-loaded solid lipid nanoparticles in-vitro condition. *Iran. J. Basic Med. Sci.* **2016**, *19*, 1231–1237. [PubMed]
131. Luque-Alcaraz, A.; Cortez-Rocha, M.; Velázquez-Contreras, C.; Acosta-Silva, A.; Santacruz-Ortega, H.; Burgos-Hernández, A.; Argüelles-Monal, W.; Plascencia-Jatomea, M. Enhanced antifungal effect of chitosan/pepper tree (*Schinus molle*) essential oil bionanocomposites on the viability of *Aspergillus parasiticus* spores. *J. Nanomater.* **2016**, *2016*, 6060137. [CrossRef]
132. Thanighaiarassu, R.R.; Sivamai, P.; Devika, R.; Nambikkairaj, B. Green synthesis of gold nanoparticles characterization by using plant essential oil menthapiperita and their antifungal activity against human pathogenic fungi. *J. Nanomed. Nanotechnol.* **2014**, *5*, 229. [CrossRef]
133. Amir, A.; Naghmeh, M. Encapsulation of Satureja khuzestanica essential oil in chitosan nanoparticles with enhanced antifungal activity. *Int. J. Nutr. Food Eng.* **2017**, *11*, 331–336. [CrossRef]
134. Kalagatur, N.; Nirmal Ghosh, O.; Sundararaj, N.; Mudili, V. Antifungal activity of chitosan nanoparticles encapsulated with *Cymbopogon martinii* essential oil on plant pathogenic fungi *Fusarium graminearum*. *Front. Pharmacol.* **2018**, *9*, 610. [CrossRef]
135. Latha, M.K.V.; Lal, A.A. Efficacy of micro algae and thyme essential oil in the management of alternaria leaf spot of broccoli (*Brassica oleracea* var. italica). *Int. J. Curr. Microbiol. Appl. Sci.* **2021**, *10*, 297–303. [CrossRef]
136. Wang, Y.; Chang, J.; Morrell, J.J.; Freitag, C.M.; Karchesy, J.J. An integrated approach using *bacillus subtilis* B26 and essential oils to limit fungal discoloration of wood. *BioResources* **2012**, *7*, 3132–3141.
137. Nissa, A.; Utami, R.; Sari, A.; Nursiwi, A. Combination effect of nisin and red ginger essential oil (*Zingiber officinale* var. rubrum) against foodborne pathogens and food spoilage microorganisms. *AIP Conf. Proc.* **2018**, *2014*, 020023. [CrossRef]

foods

MDPI

Review

Backhousia citriodora F. Muell. (Lemon Myrtle), an Unrivalled Source of Citral

Ian Southwell

Plant Science, Southern Cross University, Lismore, NSW 2480, Australia; ian.southwell@scu.edu.au

Abstract: Lemon oils are amongst the highest volume and most frequently traded of the flavor and fragrance essential oils. Citronellal and citral are considered the key components responsible for the lemon note with citral (neral + geranial) preferred. Of the myriad of sources of citral, the Australian myrtaceous tree, Lemon Myrtle, *Backhousia citriodora* F. Muell. (Myrtaceae), is considered superior. This review examines the history, the natural occurrence, the cultivation, the taxonomy, the chemistry, the biological activity, the toxicology, the standardisation and the commercialisation of *Backhousia citriodora* especially in relation to its essential oil.

Keywords: *Backhousia citriodora*; lemon myrtle; lemon oils; citral; geranial; neral; iso-citrals; citronellal; flavor; fragrance; biological activity

Citation: Southwell, I. *Backhousia citriodora* F. Muell. (Lemon Myrtle), an Unrivalled Source of Citral. *Foods* **2021**, *10*, 1596. https://doi.org/10.3390/foods10071596

Academic Editors: Verica Dragović-Uzelac and Ginés Benito Martínez-Hernández

Received: 31 May 2021
Accepted: 6 July 2021
Published: 9 July 2021

1. Introduction

There are many natural sources of lemon oil or lemon scent. According to a recent ISO Strategic Business Plan [1], the top production of lemon oils comes from lemon (7500 tonne), *Litsea cubeba* (1700 tonne), citronella (1100 tonne) and *Eucalyptus* (now *Corymbia*) *citriodora* (1000 tonne). Lemon oil itself, cold pressed from the peel of *Citrus limon* L., Rutaceae, contains 2–3% of citral (geranial + neral) [2–4], the lemon flavor ingredient. Consequently, the oil, along with numerous other citrus species, is used more for its high limonene (60–80%) and minor component content as a fragrance, health care additive [5] or solvent rather than a citral lemon flavor. Citral- and citronellal-rich oils are the commercial lemon-scented oils. Significant sources [6] of these essential oils are listed in Table 1 [6–44].

The aim of this review is to examine investigations into Lemon Myrtle, *Backhousia citriodora* F. Muell. (Myrtaceae), a source of lemon-scented essential oil, that suggest that Lemon Myrtle is superior to other current commercial sources with respect to citral content, oil yield, organoleptic and medicinal properties.

The criteria used for selection of papers for review are so numerous that they are difficult to itemize. Little was covered prior to the classical The Essential Oil series [16,33] after which chemistry papers abounded, to be followed by more recent bioactivity, toxicology, standards and commercial communications as the industry expanded, all accessed from 'in-house libraries', electronic databases and published reference lists up until mid 2021.

2. Taxonomy

2.1. Etymology

In 1845, lemon-scented myrtle was named *Backhousia citriodora* F. Muell. by botanist Ferdinand von Mueller, the genus after the English botanist, James Backhouse and the species epithet from the distinctively strong lemon scent of the foliage [45]. The genus *Backhousia*, from the Myrtaceae family, is endemic to eastern Australia and is a close relative of the genus *Choricarpia*, with which it forms the *Backhousia* alliance [7]. The primary common name "Lemon-scented myrtle" was shortened to "lemon myrtle" for the native foods industry to market the leaf for culinary use. "Sweet Verbena Myrtle" and "Lemon Ironwood" are also common names. As *B. citriodora* has two chemoypes, distinction needs to be made between the citral chemotype and the L-citronellal chemotype [38,39].

2.2. Habit

Mature lemon myrtle trees reach 8 m (25 ft) in height, or higher (to 30 m) when crowned, but are often smaller. The leaves are evergreen, opposite, lanceolate, 4–15 cm in length and 1–5 cm (0.59–0.98 in) broad, glossy green, with an entire margin. The flowers are creamy white, 5–7 mm (0.20–0.28 in) in diameter, produced in clusters at the ends of the branches from summer through to autumn, and after petal fall, the calyx is persistent [45], as shown in Figure 1.

Figure 1. *Backhousia citriodora* in plantation and in flower, showing natural distribution.

2.3. Distribution

B. citriodora is endemic to only the east coast of Australia in Queensland from the Sunshine Coast regions of Eumundi, Maroochydore, Noosa and Woondum, to the ranges west of Miriam Vale and the Mackay, Whitsunday, Townsville and Herberton regions. Plantations have been established from north Queensland to northern New South Wales for both the production of dried leaf and lemon essential oil [45]. The largest of these cover 200 and 70 acres, producing over 2400 tonne of fresh leaf on stem per annum [46].

2.4. Chemotypes

Backhousia citriodora has two essential oil chemotypes: the citral chemotype is more prevalent and is cultivated in Australia for flavoring and essential oil [45]. Citral as an isolate in steam-distilled lemon myrtle oil is typically 90–98%, and oil yield 1–3% from fresh leaf [10,20,45]. This is the highest-content natural source of citral (Table 1). The citronellal chemotype is uncommon and can be used as an insect repellent [5,44] as it has similarities to citronella (*Cymbopogon nardus*) and lemon-scented gum (*Corymbia citriodora*, formerly *Eucalyptus citriodora*). Although first reported by Penfold et al. in 1950 [38], it was only in 2001 that this chemotype was rediscovered and the oil fully characterized [39]. The unique characteristic of this chemotype is that the oil is a source of L-citronellal, whereas many sources contain either the racemic form or the D-isomer. This chemotype does not breed true as seed collected from a citronellal type tree has given progeny with a 1.05:1 ratio of the citronellal:citral chemotypes [39,45].

Table 1. Commercial and potentially commercial sources of citral (neral + geranial) and citronellal essential oil.

Lemon Constituent	Species	Common Name	Plant Part	% Oil	% Lemon Constituent	Reference
Citral (neral + geranial)	*Backhousia citriodora*	Lemon myrtle	leaves	1.1–3.2	80–97	[6–11]
	Litsea citrata				90	[11]
	Cymbopogon flexuosus	Lemongrass	leaves	0.2–0.4	60–90	[11–14]
	Cymbopogon citratus	Lemongrass West Indian	leaves	0.2–0.3	73–90	[6,15–17]
	Leptospermum liversidgei var. A.	Olive tea tree	aerial parts	0.6–0.8	70–80	[11,18–21]
	Leptospermum petersonii	Lemon tea tree	leaf	2.0–7.0	50–77	[18–21]
	Litsea cubeba	Litsea, may chang	fruit		63–78	[11,22,23]
	Aloysia triphylla (Lippia citriodora)	Lemon verbena			43–68	[6,24,25]
	Melaleuca teretifolia	Banbar or marsh honey myrtle	leaves, stems	1.5	66–68	[26,27]
	Ocimum gratissimum				66.5	[11]
	Lindera citriodora				65	[11]
	Melissa officinalis	Melissa			64	[23,28–30]
	Calypranthes parriculata				62	[11]
	Citrus limon	Lemon petitgrain	leaves and twig	0.6	7–50	[31–33]
	Ocimum x africanum	Lemon basil			42	[6,28]
	Melaleuca stipitata	Bukbuluk	Leaves	0.7–3.1	44	[34]
	Eucalyptus staigeriana	Lemon ironbark	leaves	2.9–3.4	30–50	[20,28]
	Citrus aurantifolia leaves	Petitgrain			36	[11]
	Melaleuca citrolens	Gulbarn	leaves, stems	1.3–3.9	16–43	[34–36]
	Thymus citriodorus	Lemon thyme		0.4	16	[6,37]
Citronellal	*Backhousia citriodora*	Lemon myrtle	leaves	1.8–3.2	80 -89	[20,38,39]
	Corymbia citriodora (Eucalyptus citriodora)	Lemon-scented gum	leaves and twig	0.5–4.2	65–85	[20,40,41]
	Leptospermum liversidgii Var. B	Olive tea tree	aerial parts	0.5	70–80	[20]
	Ochrosperma citriodorum (Baeckea citriodora)		aerial parts	0.3–1.0	54–80	[20]
	Leptospermum petersonii	Lemon tea tree	leaf	2.0–7.0	40	[19–21]
	Cymbopogon winterianus	Citronella, Java type	leaves	~0.5	31–40	[15,42]
	Cymbopogon nardus	Citronella, Sri Lanka type	leaves	1–9	1–47	[43,44]

2.5. Agronomy

The silvicultural and agronomic aspects of this species, including plantation development, propagation, planting and tending, growth, pests, predators, diseases, harvesting and processing are very much dependent on the individual producer. One producer has detailed his approach [45]. The trees grow best near their natural habitat. There is an increasing demand for organic oil, hence using only organically approved pesticides, herbicides and fertilizers is recommended. The tree loves water but does not like wet feet. It is frost intolerant, with two or more nights below 0 degrees deadly to seedlings. They enjoy morning sun with growth aided by windbreaks. The species responds well to nitrogen but excessive fertilizer leads to top-heavy plants, poor tree root structure and low leaf oil quality. Ornamental trees grow well in cooler climates as a shrub rather than a tree.

As *B. citriodora* is a tropical to subtropical rainforest tree, leaf production is reduced outside these natural environments. Irrigation is essential for the first years of establishment. Plantation rows need to allow for mechanized tending and harvesting with soft footprints to prevent root damage and spacing to allow for considerable foliage spread [45].

Suitable soils should be well drained and permeable, with a moderately acidic pH, with lime not recommended. Deep ripping the soil for plantation establishment allows for aeration and moisture retention. Mulching will retain moisture and reduce erosion. Planting on mounds is not recommended. Pests and diseases vary with location and climate but always need monitoring [45].

3. Uses

The organoleptic and bioactivity properties of citral have led to the essential oil of *B. citriodora* being used in a number of applications. Commercial production has two main applications [45]: fresh or dried herb sales and distillation for essential oil production. Chief secondary uses include use by florists and the plant nurseries, where flowers and leafy branches are very popular ornamentally and the tree itself is an asset to any garden.

Citral itself has a generally recognized as safe (GRAS) listing by the United States Food and Drug Administration (FDA), whereby when added to food, it is considered safe by experts [6]. Hence, lemon myrtle oil has been used for citral applications and added as a flavoring and scenting agent to foods, cosmetics, aromatherapy massage oils and various household products (such as detergents, soaps, air fresheners, and insect repellents) to give a lemon or verbena scent [6]. Citral is also an excellent starting material for the synthesis of vitamin A and the valuable fragrant ionones [6,10,20,45]. Additionally, citral has proven bioactivity for numerous potential applications [6,11,47,48] and *B. citriodora* oil or extract has been reported to possess antimicrobial [47–51], food pathogenic [52,53], postharvest pathogenic [54], skin infection [55,56] and anti-inflammatory and antioxidative [57,58] properties. Some of these will be detailed later in this review.

4. Essential Oil

The citral chemotype yields 1.1–3.2% (fresh weight of leaf) of oil with 80–98% citral [10,59]. For commercial equipment, consistent yields of 1.5% (*w*/*w*, containing some twig) were reported compared with a variable 0.4–3.2% for laboratory distillations [45].

A first report of the less common citronellal chemotype indicated yields of 0.5–0.9% (fresh weight) of oil with 62–80% citronellal [20,39]. Year-old trees from a progeny trial, however, yielded 1.8–3.2% (dry weight) with 85–89% citronellal [39]. Propagation of seed from a single citronellal-type mature tree gave mixed progeny with an approximate 1:1 ratio of the citral and citronellal chemotypes. In contrast, progeny from two citral chemotypes gave only 3/48 of the citronellal chemotype [39]. This rarer form of L or (-) citronellal provides a starting material for the stereospecific synthesis of terpenoids used in the perfume and flavor industry [20].

5. Oil Chemistry

The major components of the leaf essential oil of *B. citriodora* are shown in Table 2, Figures 2 and 3. Initially thought to be one compound, the major component was called citral because of its lemony aroma and flavor. This terpene aldehyde was found to be a mixture of the two geometric isomers neral 9 (IUPAC Name: (2E)-3,7-dimethylocta-2,6-dienal), and geranial 10 ((2Z)-3,7-dimethylocta-2,6-dienal) also known as citral a and citral b, respectively in the ratio of 1.2–1.5, as shown in Table 2 [45].

B. citriodora components were determined by gas chromatography using flame ionisation detection (GCFID) and gas chromatography–mass spectrometry (GC–MS). The most dominant of the minor components are the iso-citrals **5**, **6**, **8**. These isomers of citral seem to always co-exist with citral and are thought to be oxidative, thermal or acid/base rearrangement artefacts of citral sourced either naturally or synthetically [10,60–62]. A published patent reported the purification of citral by fractional distillation in a controlled acidic environment (pH 3–7). This procedure reduced the formation of iso-citrals [63].

Table 2. The percentage proportion ranges for key constituents in the essential oil of the citral chemotype of *Backhousia citriodora*.

Component	Min %	Max %
β -Myrcene (**1**)	tr [a]	0.7
2.3-Dehydro-1.8 cineole (**2**)	tr [a]	0.9
6-Methyl-5-hepten-2 one (**3**)	tr [a]	2.9
Citronellal (**4**)	tr [a]	1.0
exo-Isocitral [b] (**5**)	tr [a]	2.0
Z-Isocitral [b] (**6**)	tr [a]	2.7
Linalool (**7**)	tr [a]	1.0
E-Isocitral (**8**)	tr [a]	4.3
Neral (**9**) [c]	32.0	40.9
Geranial (**10**)	44.0	60.7
Nerol (**11**) [c]	tr [a]	0.6
Geraniol (**12**)	0.5	2.5
Total citral [b]	80.0	96.0

[a] tr = traces < 0.01%. [b] Total citral is the addition of all five citral isomers. [c] On non-polar gas chromatography (GC) column stationary phases, nerol often co-elutes with neral.

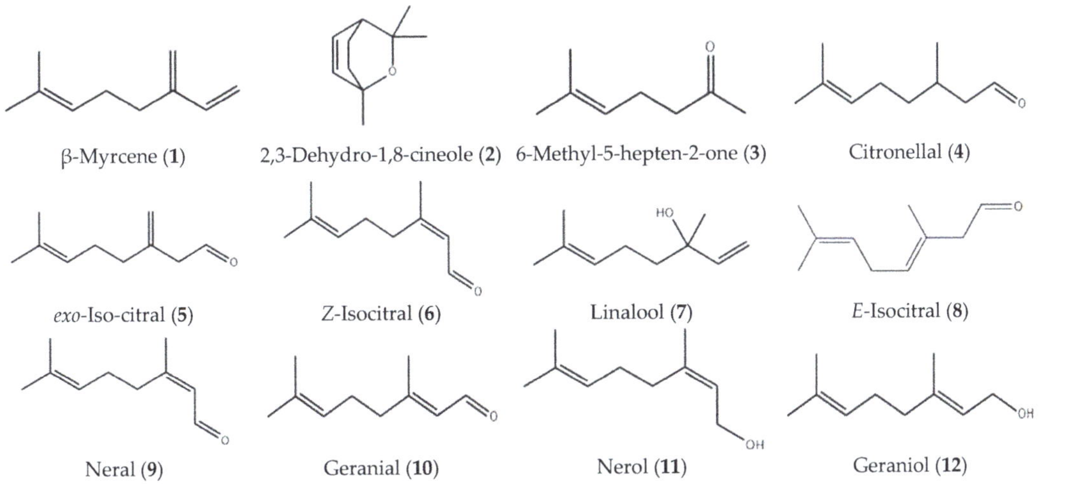

Figure 2. Major constituents of *B. citriodora* essential oil.

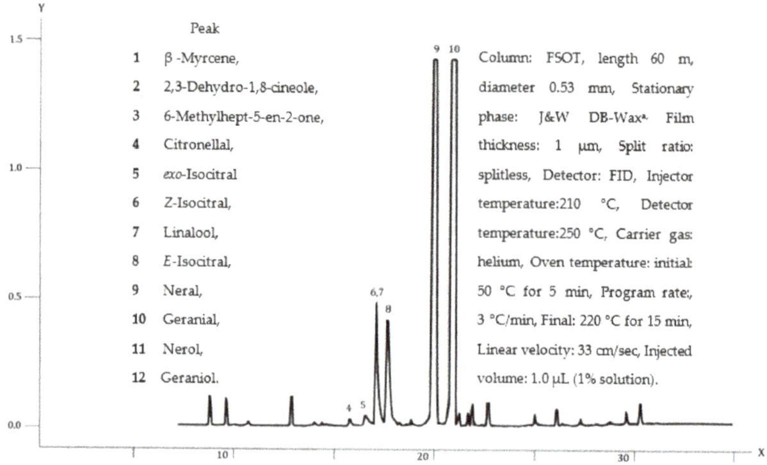

Figure 3. Gas chromatographic trace of *B. citriodora* oil on a polar column.

With gas chromatography (GC), the preferred analytical method for determining essential oil quality, the choice of solvent for injection of aldehyde-rich oils such as *B. citriodora* is important. Alcoholic or ketonic solvents such as ethanol or acetone are unsuitable because of their tendency to form acetals and ketals if left in these solvents for a length of time [64]. This was also seen in the analysis of cinnamaldehyde from Cinnamomum species using methanol as a solvent [65].

6. Bioactivity

An increasing amount of data is now being published affirming the popularity of lemon myrtle as a complimentary medicine [48,66].

Many anecdotal reports of bioactivity are now being confirmed by in vitro and in vivo investigations. The Australian Therapeutic Goods Administration (TGA) is reported to have approved three *B. citriodora* essential oil medicines by 2006 [56] and by 2017 expressed an awareness of the increasing number of products containing citral [66].

Even when Rideal–Walker co-efficients were the chief measure of microbial activity, *Backhousia citriodora* essential oil scored well [45]. Lemon myrtle oil was shown to possess significant antimicrobial activity against the organisms *Staphylococcus aureus*, *Escherichia coli*, *Pseudomonas aeruginosa*, *Candida albicans*, methicillin-resistant *S. aureus* (MRSA), *Aspergillus niger*, *Klebsiella pneumoniae* and *Propionibacterium acnes* comparable to its major component—citral [45,49–51]. For example, Minimum Inhibitory Concentrations (%v/v) against *Aspergillus niger* have been recorded as 0.1, i.e., lower than tea tree oil (0.4) and equivalent to citral (0.1) [49]. The antimicrobial activity of *B. citriodora* essential oils was found to be greater than that of citral alone and often superior to *Melaleuca alternifolia* essential oil. *B. citriodora* has significant antimicrobial activity that has potential as an antiseptic or surface disinfectant or for inclusion in foods as a natural antimicrobial agent [50].

The leaf paste has been confirmed for its antimicrobial and antifungal properties against many microbes including *Clostridium perfringens*, *Pseudomonas aeruginosa*, and a hospital isolate of methicillin-resistant *Staphylococcus aureus* (MRSA) [50,51]. Three others found the oil/extract to also be an effective antibacterial and antifungal agent against (a) food pathogenic bacteria and food spoilage yeasts [52], where damage of the yeast cell membrane through penetration caused swelling and lysis, leading to cell death; (b) against food-borne pathogens [53], where MIC values against *S. aureus* and *Escherichia coli* were 16- and 8-fold, respectively, better than tea tree oil; and (c) against the plant postharvest pathogen *Monilinia fructicola* [54], where in vitro inhibition of spore germination and mycelial growth was recorded.

Antiviral activity has been recorded in a clinical trial in treating *Molluscum contagiosum*, a skin virus causing pearly, flesh-coloured, dome-shaped papules with central umbilication frequently among children [55,56]. The trial showed that at the end of 21 days, there was a more than 90% reduction in lesions in 9/16 children treated with lemon myrtle oil.

Anti-inflammatory and antioxidative properties have also been investigated [57,58]. Lemon myrtle extract (LME) inhibited the production of inflammatory mediators such as nitric oxide (NO). Enzyme-linked immunosorbent assay and reverse-transcriptase polymerase chain reaction (RT-PCR) revealed that pretreatment with LME suppressed the protein expression and mRNA levels of pro-inflammatory cytokines such as interleukin IL-6, and tumor necrosis factor (TNF)-α in a concentration-dependent manner, respectively. This activity suggested that lemon myrtle extract could be used as a potential therapeutic agent with potent anti-inflammatory effects that could be used to treat inflammatory bowel disease. Different drying and extraction techniques for optimizing the antioxidant activity of the leaf have also been investigated [67,68].

In another study, the efficacy of lemongrass (*Cymbopogon flexuosus*) essential oil and its bioactive part citral against dual-species biofilms formed by *Staphylococcus aureus* and *Candida* species was evaluated in vitro [69]. Biofilm staining and viability tests showed both lemongrass essential oil and citral were able to reduce biofilm biomass and cell viability of each species in the biofilm.

In addition, it has been suggested that lemon myrtle extract is suitable for use in ocular health nutritional products, not because of the presence of citral in the extract, but because the extract is a source of lutein and other antioxidants along with folate and the trace minerals, magnesium and calcium [57,70].

Studies with insects have shown that effective insect repellents based on natural active ingredients can deliver repellency on par with synthetic actives in the field. For example, Greive et al. [71] showed in preliminary studies that lemon myrtle oil has insect deterrent activity. Repellency of 82% was recorded against *Aedes aegypti* mosquitoes for 30 min in laboratory tests, with greater efficacy (97%) achieved when mixed (1:5) with *Melaleuca ericifolia* oil, a source of linalool.

7. Toxicology

Citral, the major component of *Backhousia citriodora* oil, has generally recognized as safe (GRAS) status and is listed by the United States Food and Drug Administration (FDA) and hence, when added to food, is considered safe by experts [6,47,48].

When a chemical or chemical category has been agreed by the Organisation for Economic Co-operation and Development (OECD) member countries, several documents are available to the public. The OECD-generated *profile* (called either the Screening Information Dataset (SIDS) Initial Assessment Profile (SIAP) or the Initial Targeted Assessment Profile (ITAP)) contains brief summaries of SIDS endpoints as well as the major conclusions of the hazard assessment. Hence, there is much information available at sites like: https://hpvchemicals.oecd.org/UI/handler.axd?id=0ea83202-3f4f-4355-be4f-27ff02e19cb9 (accessed on 9 July 2021) [11,66,72–76] summarising the toxicology of citral. These reports draw the following conclusions:

(a) *"For human health, acute toxicity of citral was found to be low in rodents because the oral or dermal LD50 values were more than 1000 mg/kg. This chemical is irritating to skin and not irritating to eyes in rabbits, and sensitizing to skin in guinea pigs. In humans, this chemical was irritating and sensitizing to the skin at high concentrations but not by consumer products. Several repeated dose oral studies show no adverse effect of citral at less than 1000 mg/kg for 5 days to 13 weeks exposure and some histological changes in the nasal cavity or forestomach, the first exposure sites, probably due to irritation, at more than 1000 mg/kg. The NOAEL for repeat dose toxicity was 200 mg/kg/day"* and

(b) *"Citral was not carcinogenic in rats or male mice. However, there was a marginal increase in malignant lymphoma in female mice that may have been related to citral. The daily citral exposures (mg/kg/day) achieved in rats and mice at the lowest dose tested in the two-year study represents approximately 10 times the average daily intake of 5 mg/kg/day in humans"* and finally

(c) *"Under the conditions of these 2-year feed studies, there was no evidence of carcinogenic activity of citral in male or female F344/N rats exposed to 1000, 2000, or 4000 ppm. There was no evidence of carcinogenic activity of citral in male B6C3F1 mice exposed to 500, 1000, or 2000 ppm. There was equivocal evidence of carcinogenic activity in female B6C3F1 mice based on increased incidences of malignant lymphoma."* [75]

A thorough investigation of lemon balm (*Melissa officinalis* L.) essential oil has been published [30] and its oral toxicity determined in mice. Although rich in citral, a high citronellal content makes this oil more like a typical *Leptospermum petersonii* oil [18–21]. In a similar manner, *Leptospermum petersonii* was evaluated by the Complementary Medicines Evaluation Committee as an oil with citral as major component to conclude that the oil "is suitable as an excipient ingredient up to 5% concentration in Listable topical medicines only" [76].

The antimicrobial and toxicological properties of *Backhousia citriodora* essential oil, have been investigated by Hayes and Markovic, 2002 [49]. In vitro cytotoxicity testing indicated that both lemon myrtle oil and citral had a very toxic effect against human cell lines: primary cell cultures of human skin fibroblasts. However, a product containing 1% lemon myrtle oil was found to be low in toxicity and could potentially be used in the formulation of topical antimicrobial products. These same authors performed in vitro percutaneous absorption investigations of the essential oil of lemon myrtle (*B. citriodora*) on freshly excised human full-thickness abdominal skin obtained from patients undergoing elective surgery [72]. Absorption of lemon myrtle oil in human skin discs was evaluated following topical application of neat lemon myrtle oil to the epidermal surface. Citral was the only component found to be absorbing into skin at all exposure periods. When a formulated product containing 1% lemon myrtle oil was applied, total absorption of citral was measured. The histopathological assessment indicated limited damage to epidermal cells. The combination of the above methodologies enabled the generation of data that could be used for a comprehensive evaluation of the toxicity effects of lemon myrtle oil for topical application.

In a review on the "Maternal reproductive toxicity of some essential oils and their constituents", a study on citral (6) affirms *B. citriodora* as the best source of citral and specifies its non-mutagenic and non-carcinogenic attributes [72–76] and reports on an inhibition of tissue morphogenesis and tumor production. The author then reviews a host of animal studies on the reproductive toxicity of citral for animals including reduced fertility in rats, dose-dependent malformations in chicken embryos, suppression of enzymes responsible for fetal development, teratogenesis in chicken embryos and restricted fetal cranial development. One suggested action mechanism indicates competition with estrogen for estrogen receptor sites. Consequently, the use of essential oils high in citral, such as *B. citriodora*, should be restricted during pregnancy because of a possible teratogenic hazard [6].

8. Standards

Only in recent years have standards been developed for the essential oil of *Backhousia citriodora*. There have, however, been a number of monographs, especially ISO Standards, elaborated for other citral-rich [4,12,17,22,31] and citronellal-rich [40,42,43] oils.

In 2001, Standards Australia's CH21 Essential Oil Committee elaborated a monograph entitled "Oil of *Backhousia citriodora*, citral type (lemon myrtle oil)", AS 4941-2001. This Standard [8] specified appearance, colour, aroma and physical constants, i.e., specific gravity, refractive index, optical rotation, solubility in alcohol and flash point. The chromatographic table, similar to Table 2 above, listed the major components giving typical minimum and maximum percentages for each constituent. Additionally, supplied are typical chromatograms usually run on both a polar (similar to Figure 3 above) and non-polar stationary phase with significant peaks identified. Included in a 2011 amendment in this first Standard's trace were the regions where one would expect the alkanals n-octan-1-al, n-nonan-1-al, and n-decan-1-al, byproducts of the synthesis of citral to elute. Peaks in this region would indicate adulteration of the oil. This revised Standard was improved with a revision [8,77] of the geraniol percentage figures to 0.5–2.5%. This was achieved by examining the oil on gas chromatographic traces giving clear separation of geraniol and geranial which are difficult to resolve on many non-polar and intermediate-polarity stationary phases.

Approaches to the International Standard's Organisation's TC54 Essential Oil Committee in 2018 resulted in the adoption of a slightly modified version of this Australian Standard as an International Standard, which is expected to be published in 2022 [9].

9. Commerce

Although all parts of the tree, including the flowers, timber and, indeed the whole tree, can be used [45], it is the leaf that is most sought after and the main reason for plantation establishment. The leaf and terminal branches are steam distilled for a citral-rich oil used as a lemon flavor, fragrance and aromatherapy oil component. The leaf, processed as whole fresh leaf, whole dried leaf, or dried and milled herb, is also popular for lemon herbal tea and other culinary and lemon flavor uses. Lemon myrtle finds itself in teas, breads, biscuits, cakes, cheeses, chutneys, jams, pastas and vinegars, as a flavor; soaps, cosmetics and pot pourris as a fragrance; aromatherapy oils as a fragrant therapeutic; and as an air freshener, a disinfectant and in a range of body care products. Because of toxicity investigations on major component citral, topical use at less than 1% in a topical formulation is recommended [45,48].

There have been a host of industry production and use-related publications extolling the value and benefits of lemon myrtle and its essential oil [60,78–90].

Although past production figures have been difficult to acquire, several tonnes of oil and fresh or dried leaf are produced annually in Australia from millions of trees in several hundred of hectare of plantation. At the 2003 IFEAT International Conference in Sydney, an estimated current annual production of 5–8 tonne was reported [78].

The 2012 estimates of farmgate Australian production of lemon myrtle for 2011 were 575–1100 tonne leaf and 3–8 tonne oil, with a gross value of $7–23 million with 90 per cent of oil exported, mainly to the United States and the European Union [79]. According to Biosecurity Australia [79,80], 57.4 tonne of organically certified lemon myrtle oil were exported from Australia to the European Union in 2011, virtually all to Germany. Most essential oil experts consider this a highly exaggerated figure but the importance of the species as an internationally and locally traded commodity cannot be understated. In a 2014 report summarising the industry [81] and relying on the 2011 figures [79], a leaf production figure of 838 tonne of leaf was recorded. A very recent (2020) market study [82] estimates the current state of the industry and projects growth forward to 2025. A current farm gate value of Aus $12.2 m is larger than any of the other Australian native foods and botanicals and is predicted to double in the next five years. There are more than 50 enterprises producing leaf and/or oil with three of substantial size producing approximately 250 tonne of dry leaf and approximately 8 tonne of oil with farm gate values estimated at Aus $37.50 and Aus$ 350.00, respectively [82].

10. Conclusions

Lemon myrtle, *Backhousia citriodora,* citral type, is becoming established as an unrivalled source of citral lemon whether it be in leaf or oil form. With further development, this species may well become a superior source of citral. The oil yield is higher, the citral content better and the aroma cleaner, fresher and sweeter. In tree form, harvesting becomes more problematic as they do not recover and coppice from ground-level harvesting in the same manner as tea tree (*Melaleuca alternifolia*) and blue mallee (*Eucalyptus polybractea*) will do. Leaf can be hand picked or tipped with a mechanical harvester.

The lesser known citronellal chemotype is unlikely to be developed commercially until further trials are performed despite the advantages of having an excellent source of the rarer L-enantiomer [39]. Because this chemotype does not breed true, plantation trials are still at early stages and genetic material for plantations is harder to source, immediate commercialisation is not envisaged.

The medicinal properties of the citral chemotype are being increasingly investigated as efficacy in many areas is being proven in both in vitro and in vivo research. Toxicity testing is proving that the product is generally safe when used in appropriate concentrations for most applications except for pregnant mums.

Backhousia citriodora, citral type, lemon myrtle oil, has attracted the world's attention in recent decades and is consequently assured a strong place in the flavor, fragrance and health care industries for decades to come. However, there is still much work to be performed, especially at the molecular level [91] and in detecting adulteration [92].

Funding: This research received no external funding.

Acknowledgments: The author is indebted to Gary Mazzorana for plantation information, to Murray Fagg and the Australia Plant Image Index (APII), Australian National Botanic Gardens, for the *Backhousia citriodora* flower photograph and to Gary Mazzorana and Joseph Brophy for comments on the manuscript.

Conflicts of Interest: The authors declare no conflict of interest.

References

1. International Standards Organisation. *Yearly World Production of the Most Representative Essential Oils. ISO-TC54 N3114 Approval of the ISOTC 54 Strategic Business Plan*; International Standards Organisation: Geneva, Switzerland, 2019.
2. Boelens, M.H.; Jimenez, R. The Chemical Composition of Mediterranean Citrus Oils. *J. Essent. Oil Res.* **1989**, *7*, 151–159. [CrossRef]
3. Dellacassa, E.; Rossini, C.; Lorenzo, D.; Moyna, P.; Verzera, A.; Trozzi, A.; Dugo, G. Uruguayan Essential Oils. Part III. Composition of the Volatile Fraction of Lemon Essential Oil. *J. Essent. Oil Res.* **1995**, *7*, 25–37. [CrossRef]
4. International Standards Organisation. *Oil of Lemon [Citrus limon (L.) Burm. f.], Obtained by Expression, ISO 855:2003*; International Standards Organisation: Geneva, Switzerland, 2003.
5. Dosoky, N.S.; Setzer, W.N. Biological Activities and Safety of *Citrus* spp. Essential Oils. *Int. J. Mol. Sci.* **2018**, *19*, 1966. [CrossRef] [PubMed]
6. Dosoky, N.S.; Setzer, W.N. Maternal Reproductive Toxicity of Some Essential Oils and Their Constituents. *Int. J. Mol. Sci.* **2021**, *22*, 2830. [CrossRef]
7. Brophy, J.J.; Goldsack, R.J.; Fookes, C.J.R.; Forster, P.I. Leaf oils of the genus *Backhousia* (Myrtaceae). *J. Essent. Oil Res.* **1995**, *7*, 237–254. [CrossRef]
8. Standards Australia, Australia Standard. *Oil of Backhousia citriodora, Citral Type (Lemon Myrtle Oil), AS 4941-2001*; Standards Australia: Sydney, Switzerland, 2001.
9. International Standards Organisation. *Essential Oil of Lemon Myrtle (Backhousia citriodora F. Muell.), Citral Type ISO/DIS 5093-2021*; International Standards Organisation: Geneva, Switzerland, 2021.
10. Southwell, I.A.; Russell, M.; Smith, R.L.; Archer, D.W. *Backhousia citriodora* F. Muell. (Myrtaceae), a Superior Source of Citral. *J. Essent. Oil Res.* **2000**, *12*, 735–741. [CrossRef]
11. OECD. SIDS Initial Assessment Report, CITRAL CAS N°:5392-40-5 for 13th SIAM (Switzerland, 6–9 November 2001). Available online: https://hpvchemicals.oecd.org/UI/handler.axd?id=0ea83202-3f4f-4355-be4f-27ff02e19cb9 (accessed on 8 July 2021).
12. International Standards Organisation. *Oil of Lemongrass [Cymbopogon flexuosus (Nees ex Steudel) J.F. Watson], ISO 4718:2004*; International Standards Organisation: Geneva, Switzerland, 2004.
13. Lawrence, B.M. *Essential Oils 1981–1987*; Allured Publishing: Wheaton, IL, USA, 1989.
14. Lawrence, B.M. Progress in essential oils: *Ocimum gratissimum* oil, cinnamon oil, tarragon oil, and palmarosa oil. *Perfum. Flavorist* **2002**, *27*, 42–52.
15. Chagonda, L.S.; Makanda, C.; Chalchat, J.C. Essential oils of cultivated *Cymbopogon winterianus* (Jowitt) and of *C. citratus* (DC) (Stapf) from Zimbabwe. *J. Essent. Oil Res.* **2000**, *12*, 478–480. [CrossRef]
16. Guenther, E. *The Essential Oils*; D. Van Nostrand Company, Inc.: New York, NY, USA, 1950; Volume IV.
17. International Standards Organisation. *Oil of Lemongrass (Cymbopogon citratus), ISO 3217:1974*; International Standards Organisation: Geneva, Switzerland, 1974.
18. Doimo, L. (Southern Cross University, Plant Science, Lismore, NSW, Australia). Personal communication, 2001.

19. Van Vuuren, S.F.; Docrat, Y.; Kamatou, G.P.P.; Viljoen, A.M. Essential oil composition and antimicrobial interactions of understudied tea tree species. *S. Afr. J. Bot.* **2014**, *92*, 7–14. [CrossRef]
20. Lassak, E.V.; Southwell, I.A. Essential Oil Isolates from the Aust. Flora. *Int. Flav. Food Add.* **1977**, 126–132.
21. Brophy, J.J.; Goldsack, R.J.; Bean, A.R.; Forster, P.I.; Lepsch, B.J. Leaf essential oils of the genus *Leptospermum* (Myrtaceae) in eastern Australia, Part 6. *Leptospermum polygalifolium and allies. Flav. Fragr. J.* **2000**, *15*, 271–277. [CrossRef]
22. International Standards Organisation. *Oil of Litsea cubeba (Litsea cubeba Pers.), ISO 3214:2000*; International Standards Organisation: Geneva, Switzerland, 2000.
23. Lawrence, B.M. Progress in essential oils. *Perfum. Flavorist* **1996**, *21*, 59–60.
24. Santos-Gomes, P.C.; Fernandes-Ferreira, M.; Vicente, A.M.S. Composition of the Essential Oils from Flowers and Leaves of Vervain (*Aloysia triphylla* (L'Herit.) Britton) Grown in Portugal. *J. Essent. Oil Res.* **2005**, *17*, 73–78. [CrossRef]
25. Eleni Fitsiou, E.; Gregoria Mitropoulou, G.; Katerina Spyridopoulou, K.; Vamvakias, M.; Bardouki, H.; Galanis, A.; Chlichlia, K.; Kourkoutas, Y.; Panayiotidis, M.I.; Pappa, A. Chemical Composition and Evaluation of the Biological Properties of the Essential Oil of the Dietary Phytochemical *Lippia citriodora*. *Molecules* **2018**, *23*, 123. [CrossRef] [PubMed]
26. Southwell, I.A.; Russell, M.; Smith, R.L.; Brophy, J.J.; Day, J. *Melaleuca teretifolia* Chemovars: New Australian Sources of Citral and 1,8-Cineole. *J. Essent. Oil Res.* **2003**, *15*, 339–341. [CrossRef]
27. Southwell, I.A.; Russell, M.; Smith, R.L.; Brophy, J.J.; Day, J. *Melaleuca teretifolia*, a Novel Aromatic and Medicinal Plant from Australia. *Acta Hortic.* **2005**, *677*, 79–83. [CrossRef]
28. Tisserand, R.; Young, R. *Essential Oil Safety*, 2nd ed.; Elsevier: Amsterdam, The Netherlands, 2014.
29. Lawrence, B.M. Progress in essential oils. *Perfum. Flavorist* **1999**, *24*, 56–59.
30. Stojanović, N.M.; Randjelović, P.J.; Mladenović, M.Z.; Ilić, I.R.; Petrović, V.; Stojiljković, N.; Ilić, S.; Niko, S.; Radulović, N.S. Toxic essential oils, part VI: Acute oral toxicity of lemon balm (*Melissa officinalis* L.) essential oil in BALB/c mice. *Food Chem. Toxicol.* **2019**, *133*, 110794. [CrossRef]
31. International Standards Organisation. *Oil of lemon petitgrain [Citrus limon (L.) Burm. f.], ISO 8899:2003*; International Standards Organisation: Geneva, Switzerland, 2003.
32. Lawrence, B.M. Progress in Essential Oils. *Perfum. Flavorist* **1995**, *20*, 48–49.
33. Guenther, E. *The Essential Oils*; D. Van Nostrand Company, Inc.: New York, NY, USA, 1950; Volume III.
34. Brophy, J.J. Potentially Commercial Melaleucas Chpt 16 in Tea Tree, the Genus *Melaleuca*. In *Medicinal and Aromatic Plants—Industrial Profiles*; Hardman, R., Ed.; Taylor and Francis: Amsterdam, The Netherlands, 2002; Volume 9, pp. 247–274.
35. Brophy, J.J.; Clarkson, J.R. The essential oils of four chemotypes of *Melaleuca citrolens*. *J. Proc. R. Soc. NSW* **1989**, *122*, 11–18.
36. Brophy, J.J.; Craven, L.A.; Doran, J.C. *Melaleucas: Their Botany, Essential Oils and Uses (PDF)*; Australian Centre for International Agricultural Research: Canberra, Australia, 2013; p. 120. Available online: https://www.aciar.gov.au/publication/books-and-manuals/melaleucas-their-botany-essential-oils-and-uses (accessed on 8 July 2021).
37. Horváth, G.; Szabó, L.G.; Héthelyi, É.; Lemberkovics, É. Essential Oil Composition of Three Cultivated *Thymus* Chemotypes from Hungary. *J. Essent. Oil Res.* **2006**, *18*, 315–317. [CrossRef]
38. Penfold, A.R.; Morrison, F.R.; McKern, H.H.G.; Willis, J.L.; Spies, M.C. 1948 Studies in the physiological forms of the Myrtaceae, Part II. The occurrence of a physiological form of *Backhousia citriodora* F. Muell. containing laevo-citronellal. *Aust. J. Sci.* **1950**, *13*, 27.
39. Doran, J.C.; Brophy, J.J.; Lassak, E.V.; House, A.P.N. *Backhousia citriodora* F. Muell.—Rediscovery and chemical characterization of the L-citronellal form and aspects of its breeding system. *Flavour Fragr. J.* **2001**, *16*, 325–328. [CrossRef]
40. International Standards Organisation. *Essential Oil of Corymbia citriodora (Hook.) K.D. Hill & L.A.S. Johnson (syn. Eucalyptus citriodora Hook) ISO/DIS 3044:2020*; International Standards Organisation: Geneva, Switzerland, 2020.
41. Brophy, J.J.; Southwell, I.A. Eucalyptus Chemistry Chpt 5 in *Eucalyptus*. In *Medicinal and Aromatic Plants—Industrial Profiles*; Hardman, R., Ed.; Taylor and Francis: Amsterdam, The Netherlands, 2002; Volume 22, pp. 102–160.
42. International Standards Organisation. *Essential Oil of Citronella, Java Type, ISO 3848:2016*; International Standards Organisation: Geneva, Switzerland, 2016.
43. International Standards Organisation. *Oil of Citronella, Sri Lankan Type (Cymbopogon nardus (L.) W. Watson var. lenabatu Stapf.) ISO 3849:2003*; International Standards Organisation: Geneva, Switzerland, 2003.
44. Kaur, H.; Bhardwaj, U.; Kaur, R. *Cymbopogon nardus* essential oil: A comprehensive review on its chemistry and bioactivity. *J. Essent. Oil Res.* **2021**, *33*, 205–220. [CrossRef]
45. Archer, D. *Backhousia citriodora F. Muell. Lemon Scented Myrtle: Biology, Cultivation and Exploitation*; Toona: Sutton, Australia, 2004.
46. Mazzorana, G. (Australian Native Products, The Channon, NSW, Australia). Personal communication, 2020.
47. Opdyke, D.L.J. Citral. *Food Chem. Toxicol.* **1979**, *17*, 259–266.
48. Wolters Kluwer Health. Drugs.com Herbal Database, Lemon Myrtle Uses, Benefits & Dosage. Available online: https://www.drugs.com/app/lemon-myrtle.html (accessed on 14 May 2021).
49. Hayes, A.J.; Markovic, B. Toxicity of Australian essential oil *Backhousia citriodora* (Lemon myrtle). Part 1. Antimicrobial activity and in vitro cytotoxicity. *Food Chem. Toxicol.* **2002**, *40*, 535–543. [CrossRef]
50. Wilkinson, J.M.; Hipwell, M.; Ryan, T.; Cavanagh, H.M.A. Bioactivity of *Backhousia citriodora*: Antibacterial and Antifungal Activity. *J. Agric. Food Chem.* **2003**, *51*, 76–81. [CrossRef]

51. Zouhir, A.; Jridi, T.; Nefzi, A.; Ben Hamida, J.; Sebei, K. Inhibition of methicillin-resistant *Staphylococcus aureus* (MRSA) by antimicrobial peptides (AMPs) and plant essential oils. *Pharm. Biol.* **2016**, *54*, 3136–3150. [CrossRef]
52. Alderees, F.; Mereddy, R.; Webber, D.; Nirmal, N.; Sultanbawa, Y. Mechanism of Action against Food Spoilage Yeasts and Bioactivity of *Tasmannia lanceolata*, *Backhousia citriodora* and *Syzygium anisatum* Plant Solvent Extracts. *Foods* **2018**, *7*, 179. [CrossRef]
53. Thielmann, J.; Muranyi, P.; Kazman, P. Screening essential oils for their antimicrobial activities against the foodborne pathogenic bacteria *Escherichia coli* and *Staphylococcus aureus*. *Heliyon* **2019**, *5*, e01860. [CrossRef] [PubMed]
54. Lazar-Baker, E.E.; Hetherington, S.D.; Ku, V.V.; Newman, S.M. Evaluation of commercial essential oil samples on the growth of postharvest pathogen, *Monilinia fructicola* (G. Winter) Honey. *Lett. Appl. Microbiol.* **2011**, *52*, 227–232. [CrossRef]
55. Burke, B.E.; Baillie, J.E.; Olson, R.D. Essential oil of Australian lemon myrtle (*Backhousia citriodora*) in the treatment of *Molluscum contagiosum* in children. *Biomed. Pharmacother.* **2004**, *58*, 245–247. [CrossRef] [PubMed]
56. Brinckmann, J.A. Treatment of *Molluscum contagiosum* in a Child with Sensory Defensiveness. *J. Am. Herb. Guild* **2007**, *7*, 22–25.
57. Konczak, I.; Zabaras, D.; Dunstan, M.; Aguas, P.; Roulfe, P.; Pavan, A. *RIRDC Pub. No. 09/133s Health Benefits of Australian Native Foods. An Evaluation of Health-Enhancing Compounds*; Union Offset Printing: Canberra, Australia, 2009; Available online: http://www.agrifutures.com.au/wp-content/uploads/publications/09-133.pdf (accessed on 8 July 2021).
58. Shim, S.-Y.; Kim, J.-H.; Kho, K.-H.; Lee, M. Anti-inflammatory and anti-oxidative activities of lemon myrtle (*Backhousia citriodora*) leaf extract. *Toxicol. Rep.* **2020**, *27*, 277–281. [CrossRef] [PubMed]
59. Doimo, L. Iso-Citrals and Iso-Geraniols in Lemon-Myrtle (*Backhousia citriodora* F. Muell.) Essential Oils. *J. Essent. Oil Res.* **2001**, *13*, 236–237. [CrossRef]
60. Fergeus, J. What Will Be the Next Big Oil from Australia? In Proceedings of the 1999 International Federation of Essential Oil and Aroma Traders (IFEAT), Hong Kong, China, 1–4 November 1999.
61. Guenther, E. *The Essential Oils*; Van Nostrand Company, Inc.: New York, NY, USA, 1950; Volume II.
62. Ohloff, G. Zur thermischen isomerisation von citral. *Tetrahedron Lett.* **1960**, *1*, 10–14. [CrossRef]
63. Sasser, D.E. Process for the Purification of Citral. European Patent EP 0 481 798 A2, 1991.
64. Southwell, I.; Dowell, A.; Russell, M. (Plant Science, Southern Cross University, Lismore, NSW, Australia). Personal communication, 2020.
65. He, Y.; He, F.; Zhang, Y.; Wang, F.; Zheng, X.; Dai, Z.; Ma, S. Formation of cinnamaldehyde dimethyl acetal in methanol during analysis. *J. Essent. Oil Res.* **2021**. [CrossRef]
66. Therapeutic Goods Administration, Australian Government. Australian Register of Therapeutic Goods Department of Health. December 2017. Available online: https://search.tga.gov.au/s/search.html?collection=tga-websites-web&query=Lemon+Myrtle&op=Search (accessed on 6 May 2021).
67. Saifullah, M.; McCullum, R.; McCluskey, A.; Vuong, Q. Effects of different drying methods on extractable phenolic compounds and antioxidant properties from lemon myrtle dried leaves. *Heliyon* **2019**, *5*, e03044. [CrossRef]
68. Kang, E.J.; Lee, J.K.; Park, H.R.; Kim, H.; Kim, H.S.; Park, J. Antioxidant and anti-inflammatory activities of phenolic compounds extracted from lemon myrtle (*Backhousia citriodora*) leaves at various extraction conditions. *J. Food Sci. Biotechnol.* **2020**, *29*, 1425–1432. [CrossRef]
69. Gao, S.; Liu, G.; Li, J.; Chen, J.; Li, L.; Li, Z.; Zhang, X.; Zhang, S.; Thorne, R.F.; Zhang, S. Antimicrobial Activity of Lemongrass Essential Oil (*Cymbopogon flexuosus*) and Its Active Component Citral Against Dual-Species Biofilms of *Staphylococcus aureus* and *Candida* Species. *Front. Cell Infect. Microbiol.* **2020**, *10*, 603858. [CrossRef]
70. Sawant, P. Unique Australian Native Botanical—Lemon Myrtle as a Natural Source of Ocular Health Super-Nutrient. *Eye Glaucoma Res.* **2019**. Available online: https://www.boffinaccess.com/eye-and-glaucoma-research/unique-australian-native-1-102/egr-1-102.pdf (accessed on 8 July 2021).
71. Greive, K.A.; Staton, J.A.; Miller, P.F.; Peters, B.A.; Oppenheim, V.M.J. Development of *Melaleuca* oils as effective natural-based personal insect repellents. *Austral. J. Entomol.* **2010**. [CrossRef]
72. Hayes, A.J.; Markovic, B. Toxicity of Australian essential oil *Backhousia citriodora* (lemon myrtle). Part 2. Absorption and histopathology following application to human skin. *Food Chem. Toxicol.* **2003**, *41*, 1409–1416. [CrossRef]
73. Feron, V.J.; Til, H.P.; de Vrijer, F.; Woutersen, R.A.; Cassee, F.R.; van Bladeren, P.J. Aldehydes: Occurrence, carcinogenic potential, mechanism of action and risk assessment. *Mutat. Res.* **1991**, *259*, 363–385. [CrossRef]
74. Ress, N.B.; Hailey, J.R.; Maronpot, R.R.; Bucher, J.R.; Travlos, G.S.; Haseman, D.P.; Orzech, J.D.; Johnson, J.K.; Hejtmancik, M.R. Toxicology and Carcinogenesis Studies of Microencapsulated Citral in Rats and Mice. *Toxicol. Sci.* **2003**, *71*, 198–206. [CrossRef] [PubMed]
75. National Toxicology Program. NTP toxicology and carcinogenesiss studies of citral (microencapsulated) (CAS No. 5392-40-5) in F344/N rats and B6C3F1 mice (feed studies). *Tech. Rep. Ser.* **2003**, *505*, 1–268.
76. Therapeutic Goods Administration, Complementary Medicines Evaluation Committee. Extracted Ratified Minutes, Forty-Fifth Meeting, 6. Evaluation of New Substances. 6.1 *Leptospermum petersonii* (up to 5%) oil, 23 April 2004. Australian Government. Available online: https://www.tga.gov.au/sites/default/files/cmec-minutes-45.pdf (accessed on 23 May 2021).
77. Lassak, E.V. *Revision of Backhousia citriodora Essential Oil Standard*; RIRDC Publication No. 11/137 RIRDC Project No. PRJ-005404; Union Offset Printing: Canberra, Australia, 2012.

78. McCartney, W. An Introductory Overview of the Essential Oil Industry in Australia and New Zealand: Essential Oils and Aroma Chemicals—Production and Markets. In Proceedings of the IFEAT Conference, Sydney, Australia, 2–6 November 2003; pp. 3–17.
79. Clarke, M. Australian Native Food Industry Stocktake. Rural Industries Research and Development Corporation: Canberra, 2012. Publication No. 12/066. Available online: https://www.agrifutures.com.au/product/australian-native-food-industry-stocktake/ (accessed on 9 July 2021).
80. Lyall, I. (Biosecurity Australia, Canberra, Australia). Personal communication, 2012.
81. Foster, M.; Emerging Animal and Plant Industries—Their Value to Australia. Rural Industries Research and Development Corporation 2013 Publication No.14/069 Project No. PRJ—008496. Available online: http://www.agrifutures.com.au/wp-content/uploads/publications/14-069.pdf (accessed on 9 July 2021).
82. Laurie, S. *Australian Native Foods and Botanicals–2019/20 Market Study*; Australian Native Foods and Botanicals: Malanda, Australia, 2020; pp. 32–33.
83. Southwell, I.A. Lemon Myrtle. The Essential Oil. In *Australian Rainforest Bushfoods Industry Association Newsletter 1*; Australian Native Food Industry Limited: Malanda, Australia, 1996; pp. 6–7.
84. Hess-Buschmann, S. Lemon Myrtle. In *New Crop Industries Handbook*; Salvin, S., Bourke, M., Byrne, T., Eds.; Rural Industries Research and Development Corporation: Canberra, Australia, 2004; pp. 353–357, 528–531. Available online: https://www.agrifutures.com.au/wp-content/uploads/publications/04-125.pdf (accessed on 9 July 2021).
85. Hess-Buschmann, S. Lemon Myrtle. In *New Crop Industries Handbook, Native Foods*; Salvin, S., Bourke, M., Byrne, T., Eds.; Rural Industries Research and Development Corporation: Canberra, Australia, 2008; pp. 16–20. Available online: https://www.agrifutures.com.au/wp-content/uploads/publications/08-021.pdf (accessed on 9 July 2021).
86. Australian Native Food Industry Limited (ANFIL). Rural Industries Research & Development Corporation (RIRDC) Focus on Native Foods. *Lemon Myrtle, Backhousia citriodora*. Pub. No. 14/117. Available online: https://www.agrifutures.com.au/wp-content/uploads/publications/14-117.pdf (accessed on 9 July 2021).
87. Australian Native Food & Botanicals. Lemon Myrtle. Available online: https://anfab.org.au/main.asp?_=Lemon%20Myrtle (accessed on 26 May 2021).
88. Sultanbawa, Y. Lemon Myrtle (*Backhousia citriodora*) Oils. In *Essential Oils in Food Preservation, Flavor and Safety*; Preedy, V.R., Ed.; Elsevier: Amsterdam, The Netherlands, 2016; pp. 517–520. [CrossRef]
89. Pengelly, A. Antimicrobial activity of lemon myrtle and tea tree oils. *Austral. J. Med. Herb.* **2003**, *15*, 9–10.
90. Pengelly, A. *Australian Aromatic Plants Research and Therapeutics*; Penett Botanicals: North Rothbury, Australia, 2017; pp. 10–17. Available online: https://herbalfieldschool.files.wordpress.com/2017/09/australian-aromatic-plants-therapeutic.pdf (accessed on 9 July 2021).
91. Horn, T.; Barth, A.; Rühle, M.; Haser, A.; Jurges, G.; Nick, P. Molecular diagnostics of Lemon Myrtle (*Backhousia citriodora* versus *Leptospermum citratum*). *Eur. Food Res. Technol.* **2012**, *234*, 853–861. [CrossRef]
92. Nhu-Trang, T.T.; Casabianca, H.; Grenier-Loustalot, M.F. Authenticity control of essential oils containing citronellal and citral by chiral and stable-isotope gas-chromatographic analysis. *Anal. Bioanal. Chem.* **2006**, *386*, 2141–2152. [CrossRef]

foods

MDPI

Article

Biological Properties of a Citral-Enriched Fraction of *Citrus limon* Essential Oil

Marzia Pucci [1,†], Stefania Raimondo [1,†], Chiara Zichittella [1], Vincenza Tinnirello [2], Valeria Corleone [2], Gioacchino Aiello [2], Marta Moschetti [1], Alice Conigliaro [1], Simona Fontana [1,*] and Riccardo Alessandro [1]

[1] Department of Biomedicine, Neuroscience and Advanced Diagnostics (Bi.N.D), Section of Biology and Genetics, University of Palermo, 90133 Palermo, Italy; marzia.pucci@unipa.it (M.P.); stefania.raimondo@unipa.it (S.R.); zichy95.cz@gmail.com (C.Z.); marta.moschetti@unipa.it (M.M.); alice.conigliaro@unipa.it (A.C.); riccardo.alessandro@unipa.it (R.A.)

[2] Agrumaria Corleone s.p.a., Via S. Corleone, 12-Zona Ind. Brancaccio, 90124 Palermo, Italy; vincenzatinni@gmail.com (V.T.); vacorleone@agrumariacorleone.com (V.C.); gaiello@agrumariacorleone.com (G.A.)

* Correspondence: simona.fontana@unipa.it; Tel.: +39-09123865731

† These authors contributed equally to this work.

Received: 17 August 2020; Accepted: 11 September 2020; Published: 14 September 2020

Abstract: Lemon essential oil (LEO) is a well-known flavoring agent with versatile biological activities. In the present study, we have isolated and characterized four citral-enriched fractions of winter LEO. We reported that in murine and human macrophages the pre-treatment with a mix of these fractions (Cfr-LEO) reduces the expression of the pro-inflammatory cytokines TNF-α, IL-1β, and IL-6 induced by LPS. In addition, Cfr-LEO counteracts LPS-induced oxidative stress, as shown by the increase in the GSH/GSSG ratio in comparison to cells treated with LPS alone. Overall, the results reported here encourage the application of EO fractions, enriched in citral, in the nutraceutical industry, not only for its organoleptic properties but also for its protective action against inflammation and oxidative stress.

Keywords: *Citrus limon* (L.) Osbeck; essential oil; citral; nutraceuticals; inflammation; oxidative stress

1. Introduction

Essential oils (EOs), also called "volatile odoriferous oil", are concentrated aromatic liquids of a combination of volatile compounds that can be extracted from different parts of the plant, for example, leaves, peels, barks, flowers, buds, seeds, twigs, and roots. Different isolation methods have been described, among which steam distillation [1–3].

The genus *Citrus*, one of the most important taxonomic subunits of the family *Rutaceae*, includes approximately 17 species of plants that produce some of the most cultivated fruits in the world. Among them, considering their significative content, the species most used for EO production are lemons (*Citrus limon* (L.) Osbeck), oranges (*Citrus sinensis* (L.) Osbeck), grapefruits (*Citrus paradisi* Macfad), both mandarins and tangerines (*Citrus reticulata* Blanco), and various limes (mainly *Citrus Aurantifolia* (Christm.) Swingle) [4]. The main components of the *C. limon* EO (LEO) are represented by monoterpenoids [5–7]. In particular, LEO is a complex mixture of limonene, γ-terpinene, citral, linalool, β-caryophyllene, α-pinene, and β-pinene [8].

Linalool, β-caryophyllene, and limonene possess anti-inflammatory effects [9–11]; instead, α-pinene and β-pinene exert antioxidant effects by reducing nitric oxide production [12].

Moreover, citral, a mixture of the two aldehydes geranial and neral, which represents one of the main bioactive components of lemon oil (65–85%), is known to possess various medicinal properties as inhibiting oxidant activity, nuclear factor kappa B (NF-kB) activation, and cyclooxygenase-2

(COX-2) expression. Katsukawa M. et al., showed that citral, in human macrophage-like U937 cells, induces expression of PPARα and -γ responsive genes and suppressed both LPS-induced COX-2 mRNA and protein expression in a PPARγ-dependent manner [13]. Bouzenna H. et al. showed antioxidant effects of citral in rat small intestine epithelial cells, suggesting that citral can protect against aspirin-induced oxidative stress [14]. Interestingly, citral also has a significant effect in cancer prevention and treatment [15]; for example, hepatocarcinogenesis in rats was inhibited by lemongrass oil with a high content of citral [16]. Belusamy et al. also demonstrated that citral inhibited cell viability, proliferation, and clonogenic potential of prostate cancer cells by targeting key players of fatty acid biosynthesis [17].

Furthermore, citral mainly contributes to lemon flavor and for this reason, it is usually applied in the nutraceutical industry, which today is extremely interested in finding innovative solutions to obtain plant derivates with beneficial properties, to add to their existing products like food, beverages, and cosmetics.

Increasing population aging is unfortunately associated with the occurrence of age-related and inflammatory-based disorders. The relationship between oxidative stress and chronic inflammation is well known from the literature; in particular, reactive oxygen species (ROS) disrupts cell signaling, alters the metabolism of arachidonic acid, promotes or enhances airway and systemic inflammation [18]. Among inflammatory and immune effector cell types, macrophages play a crucial role in the immune response, producing both pro-inflammatory cytokines and other inflammatory mediators [19], for example, tumor necrosis factor α (TNF-α), interleukin-1β (IL-1β), IL-6. To date, pro-inflammatory and pro-oxidant regulators are considered important targets for the development of therapeutic strategies.

Different studies focused their attention on developing inhibitors from natural resources to prevent or alleviate chronic inflammatory conditions that can be used with minimal side effects.

According to the literature, LEO has versatile therapeutic activities on the digestive apparatus and the cardiovascular, nervous, and immune systems [20–31]. To date, most studies are centered on the functional analysis of whole LEO, highlighting its antioxidant and anti-inflammatory properties specifically related to the presence of linalool [11] and limonene [20]. Less consideration has been given to the biological properties of the fractions obtained from whole oil; in fact, the few available information is mainly focused on the isolation techniques and the molecular profiling description [32–36], rather than on the functional role of the fractions [37–39].

This study aimed to evaluate the anti-inflammatory and antioxidant activity of citral-enriched fractions of *C. limon* EO (Cfr-LEO) that, for their improved aroma profile, could have great commercial importance as natural food additives. With this purpose, *C. limon* EO was obtained by cold-pressed extraction from winter fruit and the Cfr-LEO isolated by chromatographic technique. The effects of Cfr-LEO were investigated on LPS-stimulated murine and human macrophages.

2. Materials and Methods

2.1. *Purification of Enriched Fractions from* Citrus limon *L. Essential Oil*

C. limon essential oil (LEO) was recovered by cold-pressed extraction mechanical process from the peels of winter fruits at the company Agrumaria Corleone S.P.A. (Palermo, Italy).

After cold dewaxing at −20 °C for 48 h and subsequent filtration through a paper filter with 10-micron pores, LEO was fractionated by a newly developed adsorption column chromatography. Some volumes of essential oil flowed through the chromatographic column filled with a particular type of stationary phase under the following operating conditions: pressure: atmospheric; temperature: 25 °C; flow: 3 mL/min. This newly developed method allowed the collections of fractions enriched with the main aromatic compounds present in *C. limon* essential oil, based on their affinity with the stationary phase. Other technical details cannot here be reported and described since to date are covered by trade secrets.

2.2. Gas Chromatography (GC-MS and GC-FID) Analyses

The composition of volatile constituents of the essential oil was analyzed with Agilent 6890 N Network gas chromatographic (GC) equipped with Agilent 5973 mass spectrometer (Agilent Technologies Inc., Santa Clara, CA, USA) with a DB-5MS fused silica column (30 m × 0.25 mm ID, film thickness 0.25 µm, Agilent Technologies Inc., Saint Clara, CA, USA). The oven program started with an initial temperature of 75 °C held for 9 min, then the oven temperature was heated at 4 °C/min to 120 °C and after at 5 °C/min to 310 °C for 15 min. The source temperature was 230 °C, quadrupole temperature was 150 °C, injector and detector were 250 and 280 °C, respectively. The carrier gas was helium adjusted to a linear velocity of 42 cm/s. Samples were prepared by diluting in hexane in a ratio of 1:10. Samples were injected (0.2 µL) with a split ratio of 1:50. The identification of essential oil volatile components was performed by comparison of their mass spectra with a NIST MS Search and Wiley 138 mass spectral library, as well as with literature data.

The GC-FID analysis of the EO was performed with GC Agilent 7890 A (Agilent Technologies Inc., Santa Clara, CA, USA) equipped with a non-polar DB-5 (Agilent Technologies Inc., Santa Clara, CA, USA) capillary column (length: 20 m; 100 µm internal diameter; film thickness 0.1µm). The oven temperature program was set to 75 °C for 3 min, 8 °C/min to 140 °C for 0 min, 30 °C/min to 310 °C for 5 min. The carrier gas was H2, at 0.80 mL/min flow. Injector and detector temperatures were set at 280 and 350 °C, respectively. Each EO sample was prepared by diluting it in isooctane in a ratio of 1:10. The injection volume was set at 0.2 µL. The GC-FID analysis of the samples allowed to obtain the relative percentage quantity of the single components contained in the analyzed samples. The value of each analyte is expressed as the percentage area of the peak with respect to the total composition of the EO obtained from the GC-FID analysis.

2.3. Cell Culture and Treatment

The murine macrophage RAW264.7 cell line was purchased from ATCC (Manassas, VA, USA). Cells were cultured in Dulbecco's modified Eagle's medium (DMEM) supplemented with 10% Fetal Bovine Serum (FBS, Euroclone, UK), 2 mM l-glutamine, 100 U/mL penicillin, and 100 µg/mL streptomycin (Euroclone, UK). The human monocyte THP-1 cell line was purchased from ATCC (Manassas, VA, USA). Cells were cultures in RPMI-1640 medium (Euroclone, UK) supplemented with 10% FBS, 2 mM L-glutamine, 100 U/mL penicillin, and 100 µg/mL streptomycin (Euroclone, UK). THP-1 monocytes were differentiated into M0 macrophages (THP-1 M0) as previously described [40]. In particular, cells were plated at 1×10^5 cells/mL and incubated at 37 °C with 5% CO_2 for 48 h in the presence of 50 ng/mL of Phorbol 12-myristate 13-acetate (PMA, Sigma-Aldrich, Saint Luis, MO, USA); subsequently, the conditioned medium containing PMA was removed and replaced with fresh medium for 3 days for cell recovering. Cells were then treated at different time points with various concentrations of LEO or Cfr-LEO, previously diluted in a solution consisting of 95% FBS and 5% DMSO. The biological assays reported in this manuscript have been carried out with 2 different batches of Cfr-LEO.

2.4. MTT (3-[4,5-Dimethylthiazol-2-yl]-2,5 Diphenyl Tetrazolium Bromide) Assay

Cell viability was determined by MTT assay as previously described [41]. RAW264.7 were seeded in triplicate at 3×10^3 cells per well in 96-well plates; 24 h post-seeding, cells were treated for 24 and 48 h with different concentrations of LEO or Cfr-LEO (0.005%, 0.01%, 0.02%, 0.05%). THP-1 M0 were seeded in triplicate at 2×10^4 cells per well in 48-well plates and treated for 24 and 48 h with LEO or Cfr-LEO (0.005%, 0.01%, 0.02%, 0.05%). The absorbance was measured by ELISA reader at 540 nm (Microplate Reader, BioTek, Winooski, VT, USA). Values are expressed as a percentage of cell growth versus control (untreated cells).

2.5. Measurement of Cytotoxicity in Cells Exposed to Cfr-LEO Treatment

For the detection of cytotoxicity, the CellTox™ Green Cytotoxicity Assay (Catalog number G8741, Promega, Madison, WI, USA) was used. RAW264.7 were cultured in triplicate at 5×10^3 cells per well into white-walled, opaque assay 96 well plates; 24h post-seeding, cells were treated for 24 and 48 h with different concentrations of LEO or Cfr-LEO (0.005%, 0.01%, 0.02%, 0.05%). THP-1 M0 cells were seeded in triplicate at 1×10^4 cells per well into white-walled, opaque assay 96 well plates and treated for 24 and 48 h with LEO or Cfr-LEO (0.005%, 0.01%, 0.02%, 0.05%). Changes in membrane integrity that occur as a result of cell death were measured by the CellTox™ Green Cytotoxicity Assay following the manufactures instructions. The fluorescence, proportional to cytotoxicity, was measured by Glomax (Promega).

2.6. RNA Isolation and Real-Time PCR

Levels of IL1β, IL6, and TNFα were measured by Real-time PCR. RAW264.7 cells were seeded at 5×10^4 cells per well in 12-well plates; 24 h after seeding, cells were treated for 2 h with different concentrations of LEO or Cfr-LEO (0.005%, 0.01%, 0.02%) and then exposed to LPS (500 ng/mL) for 6 h, without oil removal. THP-1 M0 cells were seeded at 1×10^5 cells per well in 12-well plates and treated for 2 h with LEO or Cfr-LEO (0.005%, 0.01%) before their exposure to LPS (1 µg/mL) for 6 h, without oil removal. Levels of BAX, BAD, and BCL-2 were measured by Real-Time PCR in RAW264.7 cells treated with 0.01% and 0.02% of Cfr-LEO for 24 and 48 h. RNA was isolated using the commercially available Illustra RNA spin Mini Isolation Kit (GE Healthcare, Little Chalfont, Buckinghamshire, UK), according to manufacturer's instructions. Total RNA from RAW264.7 or human THP-1 M0 cells was reverse transcribed to cDNA using the High Capacity cDNA Reverse Transcription kit (Applied Biosystems, Foster City, CA, USA). For quantitative SYBR Green Real-time PCR, the reaction was carried out in a total volume of 20 µL containing 2X SYBR Green I Master Mix (Applied Biosystems), 2 µL of cDNA, and 300 nM forward and reverse primers. The oligonucleotides used are reported in the Table 1.

Table 1. Oligonucleotides used in Real-Time PCR analysis

Gene	Forward	Reverse
Murine		
GAPDH	CCCAGAAGACTGTGGATGG	CAGATTGGGGGTAGGAACAC
BCL-2	GGACTTGAAGTGCCATTGGT	AGCCCCTCTGTGACAGCTTA
BAX	CTGCAGAGGATGATTGCTGA	GATCAGCTCGGGCACTTTAG
BAD	GAGTCGCCACAGTTCGTACC	GGTCCCATCGCACCTAACG
IL6	GAGGATACCACTCCCAACAGACC	AAGTGCATCATCGTTGTTCATACA
TNFα	CACGTCGTAGCAAACCACCAAGTGGA	TGGGAGTAGACAAGGTACAACCC
IL1β	CAACCAACAAGTGATATTCTCCATG	GATCCAACACTCTCCAGCTGCA
Human		
GAPDH	ATGGGGAAGGTGAAGGTCG	GGGTCATTGATGGCAACAATAT
IL6	GGTACATCCTCGACGGCATCT	GTGCCTCTTTGCTGCTTTCAC
TNFα	CCAGGCAGTCAGATCATCTTCTC	AGCTGGTTATCTCTCAGCTCCAC
IL1β	ACAGATGAAGTGCTCCTTCCA	GTCGGAGATTCGTAGCTGGAT

Real-time PCR was performed in 48-well plates using the Step-One Real-Time PCR System (Applied Biosystems). Relative changes in gene expression between control and treated samples were determined using the ΔΔCt method. Levels of the target transcript were normalized to a GAPDH endogenous control, constantly expressed in all samples (ΔCt). For ΔΔCt values, additional subtractions were performed between treated samples and control ΔCt values. Final values were expressed as fold change.

2.7. Enzyme Linked ImmunoSorbent Assay (ELISA) Assays

The amounts of IL6 and TNFα in culture supernatants were determined by using mouse IL6- and TNFα- specific ELISA kits (Thermo Fisher Scientific, Waltham, MA USA).

RAW264.7 cells were seeded at 5×10^4 cells per well in 12-well plates; 24 h after seeding, cells were treated for 2 h with 0.01% of Cfr-LEO and then exposed to LPS (500 ng/mL) for 6 h, without oil removal. At the end of the experimental time, the conditioned medium was collected and centrifuged to remove cellular debris. The ELISA assays were then performed according to the manufacturer's instructions.

2.8. Measurement of GSH and GSSG in Cells Exposed to Cfr-LEO Treatment

To detect and quantify total glutathione (GSH + GSSG), GSSG and GSH-to-GSSG ratios in THP-1 M0 cells treated with Cfr-LEO, the bioluminescent GSH/GSSG-Glo Assay kit (Catalog number V6611, Promega, Madison, WI, USA) was used. Cells were seeded at 1 × 104 cells per well into white-walled, opaque assay 96 well plates, treated for 2 h with Cfr-LEO (0.005% and 0.01%), and then exposed to LPS (1 μg/mL) for 6 h, without oil removal. After incubation, the bioluminescent GSH/GSSG-Glo Assay kit was used following the manufacturer's instructions. Cells were lysed by shaking with an equal volume of either total or oxidized glutathione reagent for 5 min, and then, cell lysates were incubated for 30 min with 50 μL of luciferin generation reagent. After the addition of 100 μL of luciferin detection reagent, plates were equilibrated for 15 min at room temperature. The luminescence (net relative luminescence unit, RLU) was measured by Glomax (Promega). The results were expressed as the GSH/GSSG ratio, calculated as follows: GSH/GSSG ratio = (Net total glutathione RLU−Net GSSG RLU)/(Net GSSG RLU/2).

2.9. Statistical Analysis

Data are represented as means ± SD. Comparisons were made using a Student's *t*-test. Values were considered statistically significant when $p \leq 0.05$.

3. Results and Discussion

3.1. Characterization of the Fractions Isolated from Citrus limon EO

To exploit at best the use of LEO as a flavoring agent, an aim is to obtain fractions enriched in compounds responsible for the lemon aroma. With this purpose, within this study, starting from the whole LEO extracted from winter fruits, we isolated four citral-enriched fractions with an enhanced lemon flavor.

Overall, 43 compounds, representing more than 99% of the total volatiles, were analyzed by GC-FID in the whole LEO and in each fraction (Fr). These compounds are listed in Table 2 where the corresponding Retention time and the % are reported for whole LEO and for 4 out of the 16 isolated fractions, Fr13, Fr 14, Fr15, and Fr16. These fractions were selected since they were the only in which neral and geranial (n 26 and 28 in Table 2), the two isomers of citral considered the essence of the lemon aroma, were enriched.

Table 3 summarizes the percentage trend of the main chemical classes of EOs (Monoterpenes, Aldehydes, Esters, Sesquiterpenes, Aliphatic Alcohols, Sesquiterpene Alcohols). Reported data show that within the class of aldehydes, the citral (a mixture of geranial and neral) is the major constituent and has an increment of about 30% in each of the four selected fractions in comparison to whole LEO; no differences were revealed for Monoterpenes, Sesquiterpenes, Aliphatic Alcohols, and Sesquiterpene Alcohols while there was a decrement of Esters. Internal olfactory panel tests have demonstrated that the fractions Fr13, Fr14, Fr15, and Fr16 enhanced their aroma properties in comparison to whole LEO, thus indicating that the 30% increase of citral determinates better olfactory characteristics.

Table 2. Volatile compounds identified in winter lemon essential oil (EO) (whole LEO) and in the selected fractions (Fr13, Fr14, Fr15, Fr16).

No	VOLATILE COMPOUND	WHOLE LEO		Fr13		Fr14		Fr15		Fr16	
		Rt [a]	Area (%) [b]	Rt	Area (%)	Rt	Area (%)	Rt	Area (%)	Rt	Area (%)
1	α-Thujene	1.688	0.43	1.674	0.42	1.676	0.42	1.675	0.42	1.675	0.42
2	α-Pinene	1.757	1.81	1.743	1.79	1.745	1.79	1.744	1.79	1.744	1.79
3	Camphene	1.892	0.06	1.875	0.05	1.878	0.06	1.876	0.05	1.877	0.05
4	Sabinene	2.132	1.83	2.112	1.81	2.115	1.81	2.113	1.81	2.114	1.81
5	β-Pinene	2.197	10.88	2.178	10.81	2.181	10.80	2.18	10.80	2.18	10.79
6	Myrcene	2.309	1.57	2.286	1.56	2.29	1.56	2.288	1.56	2.289	1.56
7	Octanal	2.447	0.07	2.419	0.06	2.422	0.06	2.419	0.06	2.421	0.06
8	Phellandrene	2.504	0.05	2.477	0.04	2.481	0.04	2.478	0.04	2.48	0.04
9	δ-3-Carene	2.593	0.01	2.566	0.01	2.572	0.01	2.57	0.01	2.571	0.01
10	α-Terpinene	2.683	0.21	2.654	0.20	2.661	0.20	2.657	0.20	2.657	0.20
11	Limonene	2.982	67.10	2.965	67.09	2.972	67.09	2.971	67.07	2.972	67.05
12	(Z)- β-Ocimene	3.063	0.05	3.008	0.05	3.014	0.05	3.012	0.05	3.013	0.05
13	(E)- β-Ocimene	3.198	0.12	3.163	0.12	3.17	0.12	3.166	0.12	3.168	0.12
14	γ-Terpinene	3.434	9.32	3.404	9.27	3.411	9.27	3.408	9.26	3.41	9.26
15	Cis-Sabinene hydrate	3.556	0.04	3.522	0.01	3.52	0.01	3.513	0.02	3.515	0.03
16	Octanol	3.654	0.01	3.612	0.00	3.609	0.00	3.606	0.00	3.608	0.00
17	Terpinolene	3.936	0.41	3.893	0.41	3.9	0.40	3.895	0.40	3.898	0.40
18	Trans-Sabinene hydrate	4.132	0.03	4.1	-	4.1	-	4.1	-	4.1	-
19	Linalool	4.158	0.13	4.109	0.13	4.115	0.14	4.109	0.15	4.112	0.15
20	Nonanal	4.228	0.11	4.179	0.11	4.185	0.10	4.18	0.10	4.183	0.10
21	Citronellal	5.186	0.10	5.134	0.10	5.14	0.10	5.134	0.10	5.137	0.10
22	Terpinen-4-ol	5.65	0.03	5.596	0.03	5.601	0.03	5.594	0.03	5.598	0.03
23	α-Terpineol	5.917	0.17	5.724	-	5.729	0.01	5.863	0.03	5.865	0.05
24	Decanal	6.223	0.04	6.171	0.03	6.177	0.03	6.17	0.03	6.174	0.03
25	Nerol+Citronellol	6.678	0.05	6.633	0.00	6.629	0.01	6.622	0.02	6.625	0.03
26	Neral	6.905	1.00	6.854	1.35	6.86	1.33	6.854	1.31	6.858	1.31
27	Geraniol	7.192	0.05	7.14	-	7.14	-	7.148	-	7.147	0.01
28	Geranial	7.483	1.72	7.435	2.19	7.442	2.21	7.437	2.21	7.44	2.21
29	Undecanal	8.15	0.02	8.096	0.02	8.102	0.02	8.096	0.02	8.1	0.02
30	Methyl geranate	8.682	0.02	8.628	0.03	8.634	0.03	8.628	0.03	8.632	0.03
31	Citronellyl acetate	9.006	0.03	8.951	0.03	8.956	0.03	8.951	0.03	8.955	0.03

Table 2. *Cont.*

No	VOLATILE COMPOUND	WHOLE LEO		Fr13		Fr14		Fr15		Fr16	
		Rt [a]	Area (%) [b]	Rt	Area (%)	Rt	Area (%)	Rt	Area (%)	Rt	Area (%)
32	Neryl acetate	9.203	0.46	9.15	0.51	9.155	0.49	9.15	0.49	9.154	0.48
33	Geranyl acetate	9.54	0.31	9.486	0.35	9.491	0.33	9.486	0.33	9.49	0.33
34	β-Caryophyllene	10.105	0.23	10.05	0.23	10.056	0.23	10.051	0.23	10.055	0.23
35	Bergamotene	10.409	0.31	10.357	0.32	10.362	0.31	10.357	0.32	10.361	0.32
36	Valencene	11.325	0.02	11.275	0.02	11.279	0.02	11.276	0.02	11.279	0.02
37	Bicyclogermacrene	11.374	0.05	11.325	0.04	11.328	0.05	11.325	0.04	11.328	0.04
38	(Z)-α-Bisabolene	11.475	0.05	11.432	0.04	11.435	0.04	11.432	0.04	11.435	0.04
39	β-Bisabolene	11.553	0.48	11.512	0.48	11.516	0.48	11.513	0.48	11.515	0.48
40	γ-Elemene	11.931	0.02	11.896	0.02	11.899	0.02	11.896	0.02	11.898	0.02
41	2-Norbornarol	12.859	0.02	12.83	0.02	12.832	0.02	12.831	0.02	12.831	0.02
42	Campherenol	12.947	0.02	12.919	0.02	12.92	0.02	12.919	0.02	12.92	0.03
43	Bisabolol	13.046	0.02	13.02	0.03	13.02	0.03	13.02	0.03	13.02	0.03

[a] Rt: Retention time in minutes [b] Area (%): percentage obtained by FID peak-area normalization. The values reported in the table are representative of one of the three replicates.

Table 3. The percentage trend of the main chemical classes of identified in winter lemon EO (whole LEO) and its fractions (Fr13, Fr14, Fr15, Fr16). Citral (in red) is more concentrated in the fractions than in the starting whole EO. The values reported in the table are representative of one of the three replicates.

Chemical Classes	Whole LEO (%)	Fr 13 (%)	Fr 14 (%)	Fr 15 (%)	Fr 16 (%)
Monoterpenes	93.9	93.65	93.64	93.61	93.6
Total Aldehydes	3.06	3.87	3.85	3.84	3.84
citral	2.73	3.54	3.53	3.52	3.52
(geranial + neral)	(1.72 + 1)	(2.19 + 1.35)	(2.21 + 1.33)	(2.21 + 1.31)	(2.21 + 1.31)
Aldehydes minus citral	0.33	0.33	0.32	0.32	0.32
Sesquiterpenes	1.16	1.15	1.15	1.15	1.15
Esters	0.82	0.92	0.88	0.87	0.87
Aliphatic Alcohols	0.43	0.17	0.20	0.24	0.28
Sesquiterpene Alcohols	0.06	0.07	0.07	0.07	0.07

We found that among the three technical replicates of the fractionation process, the percentage of compounds showing a coefficient of variation (CV%) ≤ 5 was 70% for Fr13, 72% for Fr14, 55.8% for Fr15, and 72% for Fr16 (Figure S1a) and in all fractions, CV% higher than 20% was not found for any compound. Overall, these results highlighted that the fractionation process used in this study ensures high reproducibility. Moreover, as reported in Figure S1b, among the four selected fractions, the percentage of the compounds including citral (specifically reported in Figure S1c) was comparable. Thus, according to the equivalent composition of these fractions, we decided to mix them to obtain a citral-enriched fraction of *C. limon* EO (Cfr-LEO), characterized by an increase in the aromatic level. In consideration of the interest that this enriched fraction may have in the food, beverage, and cosmetic industry, its biological effects were further evaluated.

3.2. *Evaluation of Macrophage Cell Viability after Cfr-LEO Exposure*

To obtain a comprehensive understanding of the doses to be further used, we carried out two complementary assays aimed at evaluating both cell viability and cell death.

The results in Figure 1a,b showed that cell treatment with 0.005%, 0.01%, and 0.02% of Cfr-LEO for 24 and 48 h did not affect the RAW264.7 cell viability; instead, a significant decrease in RAW264.7cell viability after treatment with 0.05% for 24 and 48 h was observed at both time points. No significant decrease in THP-1 M0 cells viability after treatment for 24 and 48 h with all concentrations of Cfr-LEO tested was observed. At the same experimental conditions, whole LEO treatment affected both murine and human macrophage cell viability in a dose and time-dependent manner (Figure 1c,d).

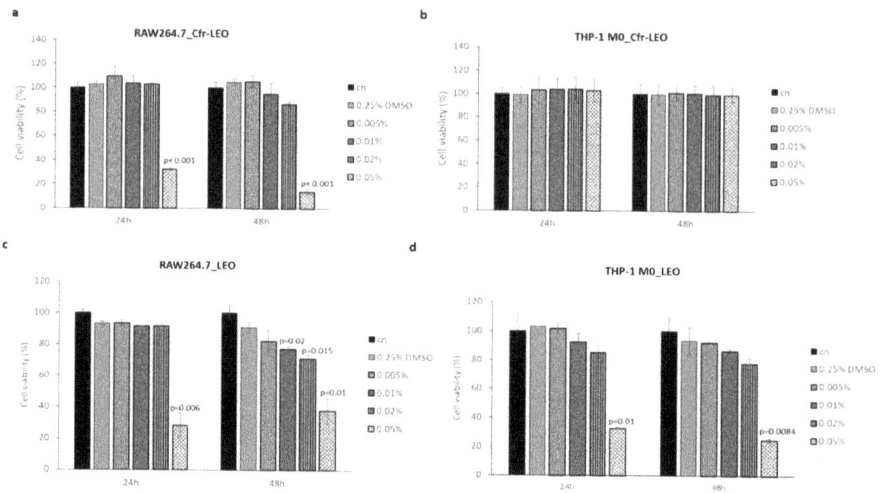

Figure 1. Evaluation of macrophage cell viability after Cfr-LEO and LEO exposure. (**a**) RAW264.7 and (**b**) THP-1 M0 cell viability was measured by MTT assay after 24 and 48 h of treatment with different concentrations of Cfr-LEO (0.005%, 0.01%, 0.02%, 0.05%). The values were plotted as the percentage of cell viability versus untreated cells (cn). Values are the mean ± SD of two biological replicates, each carried out in technical quadruplicates. The statistical significance of the differences between two groups (cells treated with 0.05% of Cfr-LEO *Vs* cn) was analyzed using a two-tailed Student's *t*-test. (**c**) RAW264.7 and (**d**) THP-1 M0 cell viability were measured by MTT assay after 24 and 48 h of treatment with different concentrations of LEO (0.005%, 0.01%, 0.02%, 0.05%). The values were plotted as the percentage of cell viability versus untreated cells (cn). Values are the mean ± SD of two biological replicates, each carried out in technical triplicates. The statistical significance of the differences between the two groups (cells treated with LEO *Vs* cn) was analyzed using a two-tailed Student's *t*-test.

Overall, the results indicate that Cfr-LEO did not exhibit any effect on cell viability at 0.005%, 0.01%, and 0.02% concentrations for both cell lines tested; interestingly, on human macrophages, the same compound can be used even at higher concentrations, up to 0.05%. Therefore, the concentrations of Cfr-LEO ranging from 0.005% to 0.02% were employed for subsequent experiments.

3.3. Evaluation of Macrophage Cytotoxicity after EO Exposure

THP-1 M0 and RAW264.7 cell lines changes in membrane integrity, that occur as a result of cell death, were measured to exclude cytotoxic effects of Cfr-LEO. Cytotoxicity of RAW264.7 cell lines after incubation for 24 and 48 h with different concentrations of Cfr-LEO (0.005%, 0.01%, 0.02%, 0.05%) was evaluated, and the obtained data are presented in Figure 2a. Data showed that treatment with 0.005%, 0.01%, and 0.02% for 24 and 48 h did not exhibit cytotoxic effects on RAW264.7 cell lines, while, consistent with cell viability assays, a significant increase in RAW264.7 cells toxicity was observed after exposure with 0.05%.

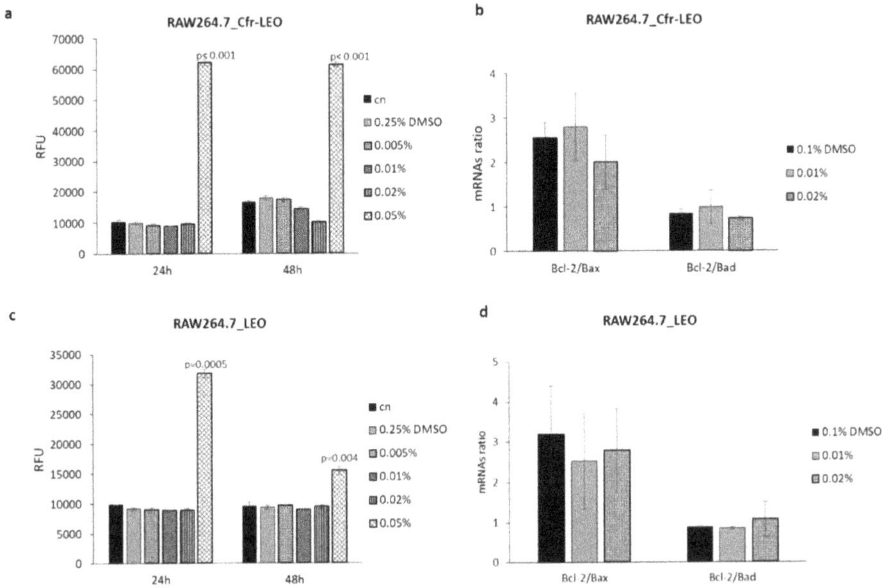

Figure 2. Evaluation of murine macrophage cytotoxicity after Cfr-LEO and LEO exposure. (**a**) RAW264.7 cell cytotoxicity was measured by CellTox Green Cytotoxicity Assay after 24 and 48 h of treatment with different concentrations of Cfr-LEO (0.005%, 0.01%, 0.02%, 0.05%). Values are plotted as Relative Fluorescence Unit (RFU). Values are the mean ± SD of two biological replicates, each carried out in technical duplicates. The statistical significance of the differences between two groups (cells treated with 0.05% of Cfr-LEO *Vs* cn) was analyzed using a two-tailed Student's *t*-test. (**b**) The effects of Cfr-LEO treatment (0.01%, 0.02%) on Bcl-2, Bax and Bad transcription were assessed by qRT-PCR. Data are represented as Bcl-2/Bax and Bcl-2/Bad ratio. Values are the mean ± SD of two biological replicates. (**c**) RAW264.7 cell cytotoxicity was measured by CellTox Green Cytotoxicity Assay after 24 and 48 h of treatment with different concentrations of LEO (0.005%, 0.01%, 0.02%, 0.05%). Values are plotted as Relative Fluorescence Unit (RFU). Values are the mean ± SD of two biological replicates, each carried out in technical triplicates. The statistical significance of the differences between two groups (cells treated with 0.05% LEO *Vs* cn) was analyzed using a two-tailed Student's *t*-test. (**d**) The effects of LEO treatment (0.01%, 0.02%) on Bcl-2, Bax, and Bad transcription were assessed by qRT-PCR. Data are represented as Bcl-2/Bax (two biological replicates) and Bcl-2/Bad (three biological replicates) ratio.

Several molecular factors such as Bcl-2, Bax, and Bad play a key role in the execution of cells apoptosis [42]. Therefore, Bcl-2, Bax, and Bad gene expression were determined at the transcriptional level in RAW264.7 cell line after incubation for 48 h with different concentrations (0.01% and 0.02%) of Cfr-LEO. In particular, apoptotic gene expression was measured based on Bcl-2/Bax and Bcl-2/Bad ratio, and the obtained data are presented in Figure 2b. No modulation in the Bcl-2/Bax and Bcl-2/Bad ratio was observed in RAW264.7 cells treated with both concentrations of Cfr-LEO tested. We observed a comparable effect of whole LEO on RAW264.7 cell toxicity induction at 0.05% treatment (Figure 2c), as well as on apoptotic gene expression level (Figure 2d).

The cytotoxic effects of Cfr-LEO and whole LEO were further evaluated on THP-1 M0 cells. Interestingly, no significant increase in THP-1 M0 cells cytotoxicity after treatment for 24 and 48 h with all concentrations of Cfr-LEO tested was observed (Figure 3a), while whole LEO treatment significantly induced THP-1 cell toxicity at 0.02% and 0.05% at both time point analyzed (Figure 3b).

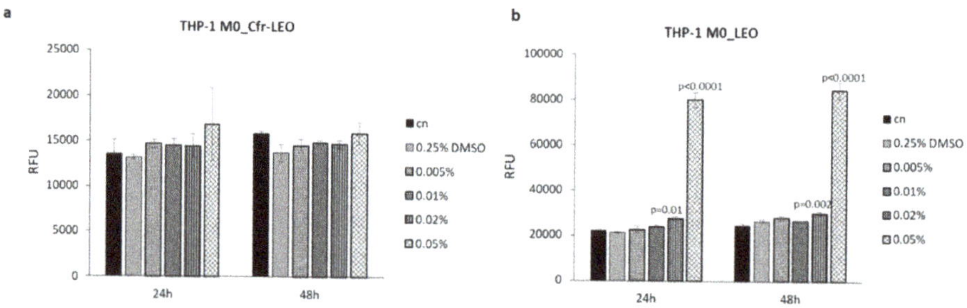

Figure 3. Evaluation of human macrophage cytotoxicity after Cfr-LEO and LEO exposure. (**a**) THP-1 M0 cell cytotoxicity was measured by CellTox Green Cytotoxicity Assay after 24 and 48 h of treatment with different concentrations of Cfr-LEO (0.005%, 0.01%, 0.02%, 0.05%). Values are plotted as the Relative Fluorescence Unit (RFU). Cn: untreated cells. Values are the mean ± SD of two biological replicates, each carried out in technical duplicates. (**b**) THP-1 M0 cell cytotoxicity was measured by CellTox Green Cytotoxicity Assay after 24 and 48 h of treatment with different concentrations of LEO (0.005%, 0.01%, 0.02%, 0.05%). Values are plotted as the Relative Fluorescence Unit (RFU). Cn: untreated cells. Values are the mean ± SD of two biological replicates, each carried out in technical triplicates.

Overall, these results, in line with the data on cell viability, confirmed that Cfr-LEO did not exhibit any cytotoxic effect at 0.005%, 0.01%, and 0.02% for both cell lines tested. Therefore, these doses were selected for further analysis.

3.4. *Protective Effects of Cfr-LEO on LPS-Activated Macrophages*

The de-regulation of inflammatory homeostasis, in favor of its over-regulation, along with the increase in the pro-inflammatory cytokines is associated with the occurrence of age-related diseases and aging itself [43–45]. This phenomenon is known as "Inflammaging" [46,47] and is correlated to the increase of TNF-α, IL-1β, and IL-6. Therefore, enhancing the anti-inflammatory response of the organism is critical for healthy aging.

Citrus EOs have largely been described for their anti-inflammatory properties [22,23,48] often related to the ability to induce a reduction of TNFα, IL1β, and IL6 levels [22,49]. *Citrus medica* EO, mostly containing limonene and γ-terpinene, reduced LPS-induced pro-inflammatory cytokines in murine macrophages; these effects occur simultaneously with the inhibition of the transcription factor NF-kB [49].

Macrophages play a key role in both specific and non-specific immune responses during inflammation [50]. Following activation by LPS, several pro-inflammatory cytokines are secreted and oxidative stress induced. Overproduction of these intermediaries is involved in different inflammatory diseases and cancer [51,52], indicating that the inhibition of macrophage activation could be an important target for inflammatory disease treatment.

To investigate whether Cfr-LEO exhibits protective effects against LPS-induced macrophage activation, we analyzed the effects of Cfr-LEO on the modulation of inflammatory mediators (IL6, IL1β, and TNFα) at the transcriptional level as well as on the GSH/GSSG ratio. In particular, as shown in Figure 4a, the pre-treatment of RAW264.7 for 2 h with each concentration of Cfr-LEO (0.005%, 0.01%, 0.02%), before their exposure to LPS for 6 h, significantly inhibited the production of pro-inflammatory mediators (IL6, IL1β, and TNFα) compared to the 6-h LPS-treated cells. Similarly, the pre-treatment of THP-1 M0 with Cfr-LEO (0.005%, 0.01%) inhibited the production of LPS-induced pro-inflammatory mediators IL6 and, TNFα, while we did not observe a reduction of IL1β Figure 4b. We observed a comparable effect of whole LEO in inhibiting the expression of inflammatory mediators in LPS-induced murine macrophages (Figure 4c), while, at the same doses whole LEO is not effective on human THP-1 macrophages (Figure 4d). The reduction of IL6 and TNFα in RAW264.7 pre-treated with 0.01% of Cfr-LEO, before exposure to LPS, was also confirmed at the protein level (Figure 4e).

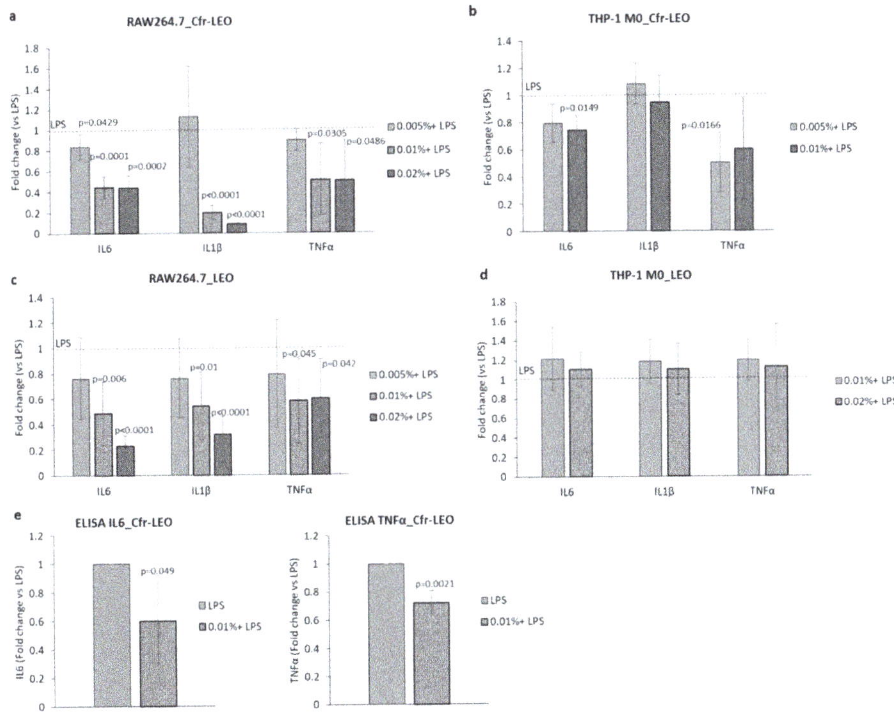

Figure 4. Anti-inflammatory effects of Cfr-LEO and LEO on LPS-activated macrophages. The anti-inflammatory effects of Cfr-LEO and LEO treatment on IL6, IL1b, and TNFα transcription levels were assessed by qRT-PCR analyses. (**a**) RAW264.7 cells were treated for 2 h with Cfr-LEO (0.005%, 0.01%, 0.02%) and then exposed to LPS for 6 h. Values are the mean ± SD of three biological replicates. (**b**) THP-1 M0 were treated for 2 h with Cfr-LEO (0.005%, 0.01%) and then exposed to LPS for 6 h. Values are reported as Fold change versus cells treated with LPS (dashed line). Values are the mean ± SD of three biological replicates for IL6 and TNFα of five biological replicates for IL1β. The statistical significance of the differences between two groups (cells treated with LPS *Vs* cells

pre-treated with Cfr-LEO + LPS) was analyzed using a two-tailed Student's *t*-test. (**c**) RAW264.7 cells were treated for 2 h with LEO (0.005%, 0.01%, 0.02%) and then exposed to LPS for 6 h. (**d**) THP-1 M0 were treated for 2 h with LEO (0.005%, 0.01%) and then exposed to LPS for 6 h. Values are reported as Fold change versus cells treated with LPS (dashed line). Values are the mean ± SD of three to six biological replicates The statistical significance of the differences between two groups (cell treated with LPS *Vs* cells pre-treated with LEO + LPS) was analyzed using a two-tailed Student's *t*-test. (**e**) IL6 and TNFα protein levels were measured by ELISA in the conditioned medium of RAW264.7 cells treated for 2 h with Cfr-LEO (0.01%) and then exposed to LPS for 6 h. Values are plotted as Fold change versus cell treated with LPS. Values are the mean ± SD of three biological replicates. The statistical significance of the differences between the two groups (cells treated with LPS *Vs* cells pre-treated with Cfr-LEO + LPS) was analyzed using a one-tailed Student's *t*-test.

Overall, these data demonstrated that the pre-treatment of macrophages with Cfr-LEO reduces the expression of the pro-inflammatory cytokines TNF-α, IL-1β, and IL-6 induced by LPS. Interestingly, since the assays we have described were performed by using both Cfr-LEO and whole LEO, we can conclude that the enrichment in citral does not alter the well-known anti-inflammatory properties of the whole LEO [22,49].

According to the literature, chronic inflammation is associated with the phenomenon of oxidative stress, caused by the imbalance between the production of reactive oxygen species and the activities of defense systems [53]. EOs are described as natural antioxidant compounds since they exert protective effects by increasing the cellular GSH levels [54] and the GSH/GSSG ratio [55]. In view of these effects, the combined administration of EOs with drugs already used in clinics may reduce the toxic effects produced by therapeutics. For example, EOs from fennel, clove, and cumin reduced the hepatotoxic effects of cyclophosphamide also by increasing GSH levels [54].

Here we also evaluated the capability of Cfr-LEO to contrast the oxidant effects of LPS in human macrophages by performing the GSH/GSSG assay. GSH/GSSG is considered a key indicator of oxidative stress [56,57]. In particular, the increase in the ratio indicates a decrease in cellular oxidative stress [58–60]. As shown in Figure 5, we found that in LPS-treated THP1 M0 cells the pre-treatment for 2 h with Cfr-LEO induced the increase of the GSH/GSSG ratio compared to no pre-treated cells, indicating the ability of Cfr-LEO to counteract the LPS-induced oxidative stress.

Overall, these results suggest that Cfr-LEO can elicit anti-inflammatory and anti-oxidant activities in LPS-stimulated macrophages, providing a rationale to consider its potential role of protective agent against inflammation stimuli.

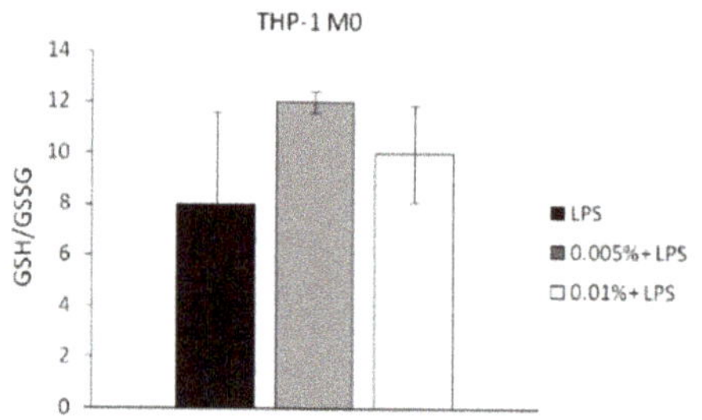

Figure 5. Anti-oxidant effect of Cfr-LEO. The anti-oxidant effects of Cfr-LEO treatment (0.005%, 0.01%) was evaluated by GSH/GSSG-Glo Assay. THP-1 M0 were treated for 2 h with Cfr-LEO (0.005%, 0.01%) and then exposed to LPS for 6 h. Values are expressed as the GSH/GSSG ratio, calculated as follows: GSH/GSSG ratio = (Net total glutathione RLU−Net GSSG RLU)/(Net GSSG RLU/2). Values are the mean ± SD of 2 independent experiments.

4. Conclusions

To date, the interest of the nutraceutical industry is to use natural compounds that possess both biological and organoleptic properties, conferred for example by citral enrichment. Therefore, in this study, we have selected and mixed four fractions with analogous composition, showing enrichment in citral, and we highlighted their beneficial properties.

In conclusion, we demonstrated that the mix of the four citral-enriched fractions of lemon essential oil (Cfr-LEO), reduces the expression of pro-inflammatory mediators and decreases oxidative stress in murine as well as human macrophage cells. Therefore, having evaluated its non-toxicity and beneficial properties, we suggest that Cfr-LEO, which can certainly be applied in the agri-food industry because of its organoleptic properties, can also represent a preventive tool for improving human health.

Supplementary Materials: The following are available online at http://www.mdpi.com/2304-8158/9/9/1290/s1, Figure S1: Evaluation of the reproducibility of the fractionation process. (**a**) The analysis of the coefficient of variation (CV %) for the three replicates. (**b**) Graph showing the percentage of each compound in the four examined fractions. Each point represents the mean ±SD of the three independent replicates of the fractionation process. (**c**) The histogram represents the mean ±SD of the percentage of citral in the selected fractions. Values are the mean of three independent replicates of the fractionation process.

Author Contributions: Experimental planning and data interpretations, S.F., M.P., S.R., and R.A.; experimental execution, M.P., S.R., C.Z., V.T., M.M., and G.A.; writing—original draft preparation, M.P. and S.R.; writing—review and editing, S.F., V.C., A.C., and R.A.; funding acquisition, V.C. and R.A. All authors have read and agreed to the published version of the manuscript.

Funding: This study was funded by a grant to Riccardo Alessandro and to Agrumaria Corleone s.p.a.: "Salute in tazza: innovative soluzioni per l'isolamento e l'applicazione industriale, anche in capsule, di composti biofunzionali ottenuti da agrumi di Sicilia" -PON I&C - MISE F/050039/01-03/X32. Marzia Pucci is supported by an AIRC fellowship (project n°22542). Stefania Raimondo is supported by PON "Ricerca e Innovazione" 2014–2020-Azione 1.2 459 "Mobilità dei Ricercatori"-AIM "Attraction and International Mobility".

Conflicts of Interest: The authors declare no conflict of interest. The funders had no role in the design of the study; in the collection, analyses, or interpretation of data; in the writing of the manuscript; or in the decision to publish the results.

References

1. Tongnuanchan, P.; Benjakul, S. Essential oils: Extraction, bioactivities, and their uses for food preservation. *J. Food Sci.* **2014**, *79*, R1231–R1249. [CrossRef]
2. Jeliazkova, E.; Zheljazkov, V.D.; Kacaniova, M.; Astatkie, T.; Tekwani, B.L. Sequential elution of essential oil constituents during steam distillation of hops (humulus lupulus l.) and influence on oil yield and antimicrobial activity. *J. Oleo Sci.* **2018**, *67*, 871–883. [CrossRef] [PubMed]
3. Celano, R.; Piccinelli, A.L.; Pagano, I.; Roscigno, G.; Campone, L.; De Falco, E.; Russo, M.; Rastrelli, L. Oil distillation wastewaters from aromatic herbs as new natural source of antioxidant compounds. *Food Res. Int.* **2017**, *99*, 298–307. [CrossRef] [PubMed]
4. Klimek-Szczykutowicz, M.; Szopa, A.; Ekiert, H. Citrus limon (lemon) phenomenon-a review of the chemistry, pharmacological properties, applications in the modern pharmaceutical, food, and cosmetics industries, and biotechnological studies. *Plants* **2020**, *9*, 119. [CrossRef]
5. Russo, M.; Bonaccorsi, I.; Costa, R.; Trozzi, A.; Dugo, P.; Mondello, L. Reduced time hplc analyses for fast quality control of citrus essential oils. *J. Essent. Oil Res.* **2015**, *27*, 307–315. [CrossRef]
6. Kaskoos, R.A. Essential oil analysis by gc-ms and analgesic activity of lippia citriodora and citrus limon. *J. Essent. Oil Bear. Plants* **2019**, *22*, 273–281. [CrossRef]
7. Gonzalez-Molina, E.; Dominguez-Perles, R.; Moreno, D.A.; Garcia-Viguera, C. Natural bioactive compounds of citrus limon for food and health. *J. Pharm. Biomed. Anal.* **2010**, *51*, 327–345. [CrossRef]
8. Gamarra, F.M.C.; Sakanaka, L.S.; Tambourgi, E.B.; Cabral, F.A. Influence on the quality of essential lemon (citrus aurantifolia) oil by distillation process. *Braz. J. Chem. Eng.* **2006**, *23*, 147–151. [CrossRef]
9. Luyendyk, J.P.; Kassel, K.M.; Allen, K.; Guo, G.L.; Li, G.; Cantor, G.H.; Copple, B.L. Fibrinogen deficiency increases liver injury and early growth response-1 (egr-1) expression in a model of chronic xenobiotic-induced cholestasis. *Am. J. Pathol.* **2011**, *178*, 1117–1125. [CrossRef]

10. Hirota, R.; Roger, N.N.; Nakamura, H.; Song, H.S.; Sawamura, M.; Suganuma, N. Anti-inflammatory effects of limonene from yuzu (citrus junos tanaka) essential oil on eosinophils. *J. Food Sci.* **2010**, *75*, H87–H92. [CrossRef]
11. Peana, A.T.; Rubattu, P.; Piga, G.G.; Fumagalli, S.; Boatto, G.; Pippia, P.; De Montis, M.G. Involvement of adenosine a1 and a2a receptors in (-)-linalool-induced antinociception. *Life Sci.* **2006**, *78*, 2471–2474. [CrossRef] [PubMed]
12. Baik, J.S.; Kim, S.S.; Lee, J.A.; Oh, T.H.; Kim, J.Y.; Lee, N.H.; Hyun, C.G. Chemical composition and biological activities of essential oils extracted from korean endemic citrus species. *J. Microbiol. Biotechnol.* **2008**, *18*, 74–79. [PubMed]
13. Katsukawa, M.; Nakata, R.; Takizawa, Y.; Hori, K.; Takahashi, S.; Inoue, H. Citral, a component of lemongrass oil, activates pparalpha and gamma and suppresses cox-2 expression. *Biochim. Biophys. Acta* **2010**, *1801*, 1214–1220. [CrossRef] [PubMed]
14. Bouzenna, H.; Hfaiedh, N.; Giroux-Metges, M.A.; Elfeki, A.; Talarmin, H. Biological properties of citral and its potential protective effects against cytotoxicity caused by aspirin in the iec-6 cells. *Biomed. Pharmacother.* **2017**, *87*, 653–660. [CrossRef] [PubMed]
15. Bhalla, Y.; Gupta, V.K.; Jaitak, V. Anticancer activity of essential oils: A review. *J. Sci. Food Agric.* **2013**, *93*, 3643–3653. [CrossRef]
16. Puatanachokchai, R.; Kishida, H.; Denda, A.; Murata, N.; Konishi, Y.; Vinitketkumnuen, U.; Nakae, D. Inhibitory effects of lemon grass (cymbopogon citratus, stapf) extract on the early phase of hepatocarcinogenesis after initiation with diethylnitrosamine in male fischer 344 rats. *Cancer Lett.* **2002**, *183*, 9–15. [CrossRef]
17. Balusamy, S.R.; Perumalsamy, H.; Veerappan, K.; Huq, M.A.; Rajeshkumar, S.; Lakshmi, T.; Kim, Y.J. Citral induced apoptosis through modulation of key genes involved in fatty acid biosynthesis in human prostate cancer cells: In silico and in vitro study. *Biomed. Res. Int.* **2020**, *2020*, 6040727. [CrossRef]
18. Moreno-Macias, H.; Romieu, I. Effects of antioxidant supplements and nutrients on patients with asthma and allergies. *J. Allergy Clin. Immunol.* **2014**, *133*, 1237–1244, Quiz 1245. [CrossRef]
19. Van Dyke, T.E.; van Winkelhoff, A.J. Infection and inflammatory mechanisms. *J. Clin. Periodontol.* **2013**, *40* (Suppl. S14), S1–S7.
20. Yoon, W.J.; Lee, N.H.; Hyun, C.G. Limonene suppresses lipopolysaccharide-induced production of nitric oxide, prostaglandin e2, and pro-inflammatory cytokines in raw 264.7 macrophages. *J. Oleo Sci.* **2010**, *59*, 415–421. [CrossRef]
21. Shen, C.Y.; Jiang, J.G.; Zhu, W.; Ou-Yang, Q. Anti-inflammatory effect of essential oil from citrus aurantium l. Var. Amara engl. *J. Agric. Food Chem.* **2017**, *65*, 8586–8594. [CrossRef] [PubMed]
22. Amorim, J.L.; Simas, D.L.; Pinheiro, M.M.; Moreno, D.S.; Alviano, C.S.; da Silva, A.J.; Fernandes, P.D. Anti-inflammatory properties and chemical characterization of the essential oils of four citrus species. *PLoS ONE* **2016**, *11*, e0153643. [CrossRef] [PubMed]
23. Ben Hsouna, A.; Gargouri, M.; Dhifi, W.; Ben Saad, R.; Sayahi, N.; Mnif, W.; Saibi, W. Potential anti-inflammatory and antioxidant effects of citrus aurantium essential oil against carbon tetrachloride-mediated hepatotoxicity: A biochemical, molecular and histopathological changes in adult rats. *Environ. Toxicol.* **2019**, *34*, 388–400. [CrossRef] [PubMed]
24. Bouzenna, H.; Samout, N.; Dhibi, S.; Mbarki, S.; Akermi, S.; Khdhiri, A.; Elfeki, A.; Hfaiedh, N. Protective effect of essential oil from citrus limon against aspirin-induced toxicity in rats. *Hum. Exp. Toxicol.* **2019**, *38*, 499–509. [CrossRef]
25. Wang, L.; Zhang, Y.; Fan, G.; Ren, J.N.; Zhang, L.L.; Pan, S.Y. Effects of orange essential oil on intestinal microflora in mice. *J. Sci. Food Agric.* **2019**, *99*, 4019–4028. [CrossRef]
26. Moraes, T.M.; Rozza, A.L.; Kushima, H.; Pellizzon, C.H.; Rocha, L.R.; Hiruma-Lima, C.A. Healing actions of essential oils from citrus aurantium and d-limonene in the gastric mucosa: The roles of vegf, pcna, and cox-2 in cell proliferation. *J. Med. Food* **2013**, *16*, 1162–1167. [CrossRef]
27. Kang, P.; Suh, S.H.; Min, S.S.; Seol, G.H. The essential oil of citrus bergamia risso induces vasorelaxation of the mouse aorta by activating k(+) channels and inhibiting ca(2+) influx. *J. Pharm. Pharmacol.* **2013**, *65*, 745–749. [CrossRef]

28. Kang, P.; Ryu, K.H.; Lee, J.M.; Kim, H.K.; Seol, G.H. Endothelium- and smooth muscle-dependent vasodilator effects of citrus aurantium l. Var. Amara: Focus on ca(2+) modulation. *Biomed. Pharmacother.* **2016**, *82*, 467–471. [CrossRef]
29. Oboh, G.; Olasehinde, T.A.; Ademosun, A.O. Essential oil from lemon peels inhibit key enzymes linked to neurodegenerative conditions and pro-oxidant induced lipid peroxidation. *J. Oleo Sci.* **2014**, *63*, 373–381. [CrossRef]
30. Costa, C.A.; Cury, T.C.; Cassettari, B.O.; Takahira, R.K.; Florio, J.C.; Costa, M. Citrus aurantium l. Essential oil exhibits anxiolytic-like activity mediated by 5-ht(1a)-receptors and reduces cholesterol after repeated oral treatment. *BMC Complement. Altern. Med.* **2013**, *13*, 42. [CrossRef]
31. Zhang, L.L.; Yang, Z.Y.; Fan, G.; Ren, J.N.; Yin, K.J.; Pan, S.Y. Antidepressant-like effect of citrus sinensis (l.) osbeck essential oil and its main component limonene on mice. *J. Agric. Food Chem.* **2019**, *67*, 13817–13828. [CrossRef] [PubMed]
32. Frolova, N.; Ukrainets, A.; Sylka, I.; Nemirich, A.; Kuzmin, O. Separation of terpenes from lemon essential oil by selective fractionation under a vacuum. *East. Eur. J. Enterp. Technol.* **2019**, *2*, 32–36. [CrossRef]
33. Warsito, W.; Palungan, M.H.; Utomo, E.P. Profiling study of the major and minor components of kaffir lime oil (citrus hystrix dc.) in the fractional distillation process. *Pan Afr. Med. J.* **2017**, *27*, 282. [CrossRef] [PubMed]
34. Kapsaski-Kanelli, V.N.; Evergetis, E.; Michaelakis, A.; Papachristos, D.P.; Myrtsi, E.D.; Koulocheri, S.D.; Haroutounian, S.A. "Gold" pressed essential oil: An essay on the volatile fragment from citrus juice industry by-products chemistry and bioactivity. *Biomed. Res. Int.* **2017**, *2017*, 2761461. [CrossRef] [PubMed]
35. Mehl, F.; Marti, G.; Boccard, J.; Debrus, B.; Merle, P.; Delort, E.; Baroux, L.; Raymo, V.; Velazco, M.I.; Sommer, H.; et al. Differentiation of lemon essential oil based on volatile and non-volatile fractions with various analytical techniques: A metabolomic approach. *Food Chem.* **2014**, *143*, 325–335. [CrossRef] [PubMed]
36. Dugo, P.; Ragonese, C.; Russo, M.; Sciarrone, D.; Santi, L.; Cotroneo, A.; Mondello, L. Sicilian lemon oil: Composition of volatile and oxygen heterocyclic fractions and enantiomeric distribution of volatile components. *J. Sep. Sci.* **2010**, *33*, 3374–3385. [CrossRef]
37. Lu, Q.; Huang, N.; Peng, Y.; Zhu, C.; Pan, S. Peel oils from three citrus species: Volatile constituents, antioxidant activities and related contributions of individual components. *J. Food Sci. Technol.* **2019**, *56*, 4492–4502. [CrossRef]
38. Navarra, M.; Ferlazzo, N.; Cirmi, S.; Trapasso, E.; Bramanti, P.; Lombardo, G.E.; Minciullo, P.L.; Calapai, G.; Gangemi, S. Effects of bergamot essential oil and its extractive fractions on sh-sy5y human neuroblastoma cell growth. *J. Pharm. Pharmacol.* **2015**, *67*, 1042–1053. [CrossRef]
39. Hu, W.; Zhang, N.; Chen, H.; Zhong, B.; Yang, A.; Kuang, F.; Ouyang, Z.; Chun, J. Fumigant activity of sweet orange essential oil fractions against red imported fire ants (hymenoptera: Formicidae). *J. Econ. Entomol.* **2017**, *110*, 1556–1562. [CrossRef]
40. Shiratori, H.; Feinweber, C.; Luckhardt, S.; Linke, B.; Resch, E.; Geisslinger, G.; Weigert, A.; Parnham, M.J. Thp-1 and human peripheral blood mononuclear cell-derived macrophages differ in their capacity to polarize in vitro. *Mol. Immunol.* **2017**, *88*, 58–68. [CrossRef]
41. Raimondo, S.; Naselli, F.; Fontana, S.; Monteleone, F.; Lo Dico, A.; Saieva, L.; Zito, G.; Flugy, A.; Manno, M.; Di Bella, M.A.; et al. Citrus limon-derived nanovesicles inhibit cancer cell proliferation and suppress cml xenograft growth by inducing trail-mediated cell death. *Oncotarget* **2015**, *6*, 19514–19527. [CrossRef] [PubMed]
42. Danial, N.N. Bcl-2 family proteins: Critical checkpoints of apoptotic cell death. *Clin. Cancer Res.* **2007**, *13*, 7254–7263. [CrossRef] [PubMed]
43. Rea, I.M.; Gibson, D.S.; McGilligan, V.; McNerlan, S.E.; Alexander, H.D.; Ross, O.A. Age and age-related diseases: Role of inflammation triggers and cytokines. *Front. Immunol.* **2018**, *9*, 586. [CrossRef]
44. Chung, H.Y.; Kim, D.H.; Lee, E.K.; Chung, K.W.; Chung, S.; Lee, B.; Seo, A.Y.; Chung, J.H.; Jung, Y.S.; Im, E.; et al. Redefining chronic inflammation in aging and age-related diseases: Proposal of the senoinflammation concept. *Aging Dis.* **2019**, *10*, 367–382. [CrossRef] [PubMed]
45. Aiello, A.; Farzaneh, F.; Candore, G.; Caruso, C.; Davinelli, S.; Gambino, C.M.; Ligotti, M.E.; Zareian, N.; Accardi, G. Immunosenescence and its hallmarks: How to oppose aging strategically? A review of potential options for therapeutic intervention. *Front. Immunol.* **2019**, *10*, 2247. [CrossRef]

46. Xia, S.; Zhang, X.; Zheng, S.; Khanabdali, R.; Kalionis, B.; Wu, J.; Wan, W.; Tai, X. An update on inflamm-aging: Mechanisms, prevention, and treatment. *J. Immunol. Res.* **2016**, *2016*, 8426874. [CrossRef]
47. Franceschi, C.; Bonafe, M.; Valensin, S.; Olivieri, F.; De Luca, M.; Ottaviani, E.; De Benedictis, G. Inflamm-aging. An evolutionary perspective on immunosenescence. *Ann. NY Acad. Sci.* **2000**, *908*, 244–254. [CrossRef]
48. Plastina, P.; Apriantini, A.; Meijerink, J.; Witkamp, R.; Gabriele, B.; Fazio, A. In vitro anti-inflammatory and radical scavenging properties of chinotto (citrus myrtifolia raf.) essential oils. *Nutrients* **2018**, *10*, 783. [CrossRef]
49. Kim, K.N.; Ko, Y.J.; Yang, H.M.; Ham, Y.M.; Roh, S.W.; Jeon, Y.J.; Ahn, G.; Kang, M.C.; Yoon, W.J.; Kim, D.; et al. Anti-inflammatory effect of essential oil and its constituents from fingered citron (citrus medica l. Var. Sarcodactylis) through blocking jnk, erk and nf-kappab signaling pathways in lps-activated raw 264.7 cells. *Food Chem. Toxicol.* **2013**, *57*, 126–131. [CrossRef]
50. Kiefer, R.; Kieseier, B.C.; Stoll, G.; Hartung, H.P. The role of macrophages in immune-mediated damage to the peripheral nervous system. *Prog. Neurobiol.* **2001**, *64*, 109–127. [CrossRef]
51. Mantovani, A.; Allavena, P.; Sica, A.; Balkwill, F. Cancer-related inflammation. *Nature* **2008**, *454*, 436–444. [CrossRef] [PubMed]
52. Elinav, E.; Nowarski, R.; Thaiss, C.A.; Hu, B.; Jin, C.; Flavell, R.A. Inflammation-induced cancer: Crosstalk between tumours, immune cells and microorganisms. *Nat. Rev. Cancer* **2013**, *13*, 759–771. [CrossRef] [PubMed]
53. Cannizzo, E.S.; Clement, C.C.; Sahu, R.; Follo, C.; Santambrogio, L. Oxidative stress, inflamm-aging and immunosenescence. *J. Proteom.* **2011**, *74*, 2313–2323. [CrossRef] [PubMed]
54. Sheweita, S.A.; El-Hosseiny, L.S.; Nashashibi, M.A. Protective effects of essential oils as natural antioxidants against hepatotoxicity induced by cyclophosphamide in mice. *PLoS ONE* **2016**, *11*, e0165667. [CrossRef]
55. Porres-Martinez, M.; Gonzalez-Burgos, E.; Carretero, M.E.; Gomez-Serranillos, M.P. Protective properties of salvia lavandulifolia vahl. Essential oil against oxidative stress-induced neuronal injury. *Food Chem. Toxicol.* **2015**, *80*, 154–162. [CrossRef]
56. Ho, E.; Karimi Galougahi, K.; Liu, C.C.; Bhindi, R.; Figtree, G.A. Biological markers of oxidative stress: Applications to cardiovascular research and practice. *Redox Biol.* **2013**, *1*, 483–491. [CrossRef]
57. Valko, M.; Leibfritz, D.; Moncol, J.; Cronin, M.T.; Mazur, M.; Telser, J. Free radicals and antioxidants in normal physiological functions and human disease. *Int. J. Biochem. Cell Biol.* **2007**, *39*, 44–84. [CrossRef]
58. Balkrishna, A.; Solleti, S.K.; Singh, H.; Tomer, M.; Sharma, N.; Varshney, A. Calcio-herbal formulation, divya-swasari-ras, alleviates chronic inflammation and suppresses airway remodelling in mouse model of allergic asthma by modulating pro-inflammatory cytokine response. *Biomed. Pharmacother.* **2020**, *126*, 110063. [CrossRef]
59. Sousa, M.S.B.; Holanda, I.M.S.; Monteiro, H.M.C.; Amancio-Dos-Santos, A. Antioxidant extract counteracts the effects of aging on cortical spreading depression and oxidative stress in the brain cortex. *Acta Cir. Bras.* **2018**, *33*, 472–482. [CrossRef]
60. Ni, J.; Wang, Q.; Shah, F.A.; Liu, W.; Wang, D.; Huang, S.; Fu, S.; Wu, L. Exogenous melatonin confers cadmium tolerance by counterbalancing the hydrogen peroxide homeostasis in wheat seedlings. *Molecules* **2018**, *23*, 799. [CrossRef]

foods

MDPI

Article

Dietary Intake of Coumarins and Furocoumarins through Citrus Beverages: A Detailed Estimation by a HPLC-MS/MS Method Combined with the Linear Retention Index System

Adriana Arigò [1], Francesca Rigano [1,*], Marina Russo [1], Emanuela Trovato [2], Paola Dugo [1,2,3] and Luigi Mondello [1,2,3,4]

1 Department of Chemical, Biological, Pharmaceutical and Environmental Sciences, University of Messina, 98168 Messina, Italy; aarigo@unime.it (A.A.); marina.russo@unime.it (M.R.); pdugo@unime.it (P.D.); lmondello@unime.it (L.M.)
2 Chromaleont s.r.l., c/o Department of Chemical, Biological, Pharmaceutical and Environmental Sciences, University of Messina, 98198 Messina, Italy; emanuela.trovato@chromaleont.it
3 Unit of Food Science and Nutrition, Department of Medicine, University Campus Bio-Medico of Rome, 00128 Rome, Italy
4 BeSep s.r.l., c/o Department of Chemical, Biological, Pharmaceutical and Environmental Sciences, University of Messina, 98168 Messina, Italy
* Correspondence: francesca.rigano@unime.it

Citation: Arigò, A.; Rigano, F.; Russo, M.; Trovato, E.; Dugo, P.; Mondello, L. Dietary Intake of Coumarins and Furocoumarins through Citrus Beverages: A Detailed Estimation by a HPLC-MS/MS Method Combined with the Linear Retention Index System. *Foods* **2021**, *10*, 1533. https://doi.org/10.3390/foods10071533

Academic Editors: Ian Southwell and Oscar Núñez

Received: 31 May 2021
Accepted: 29 June 2021
Published: 2 July 2021

Abstract: Official regulations concerning the maximum number of substances in food are introduced as a consequence of possible adverse effects, after oral administration. In this regard, analytical methods are necessary in order to determine specific targets. Among oxygen heterocyclic compounds (OHCs, that are furocoumarins, coumarins and polymethoxyflavones), only coumarin is subject to restriction by the *Regulation (EC) No 1334/2008* of the European Parliament. Furocoumarins are known for their phototoxicity and other side effects due to their dietary intake; however, an official limit about the maximum content of these compounds in food is still missing. The lack of information about the real amount of these compounds in food is responsible for the conflicting opinions about the introduction of an official limit. The HPLC-MS/MS method here proposed, in combination with the linear retention index system, represents an innovative analytical strategy for the characterization of OHCs in citrus beverages. Several types of drinks were analysed in order to quantify 35 OHCs in total. This method is suitable for the quality control of OHCs in food and the obtained results may be considered as informative data useful for the regulatory authorities in the emission of new opinions and for a potential new regulation in this field.

Keywords: beverages; *Citrus*; coumarins; furocoumarins; polymethoxyflavones; linear retention indices; liquid chromatography; tandem mass spectrometry; multiple reaction monitoring; quality control

1. Introduction

Furocoumarins (FCs) are secondary plant metabolites produced in response to attack by pests and to stressful challenges. They are particularly prevalent in the Apiaceae, Umbelliferae [1,2], Fabaceae and Rutaceae [3] families. Edible fruits and vegetables may also contain FCs, e.g., parsnips (Pastinaca sativa), in which the highest content of FCs was detected [4].

Among Rutaceae botanical species, FCs are characteristic compounds of citrus peel [5]. Recently, 61 citrus species, chosen as representatives of the genetic diversity of the species, were analysed to determine the contents of these compounds in peels and pulps [6]. Among them, a large amount of FCs was found in grapefruit (*Citrus Paradisi*), where they are responsible for the so called "grapefruit juice effect", which consists of the dangerous interaction with drug metabolism, specifically in the inhibition of the enzymes involved in the catabolism of some medications; thus, enhancing their activity and causing serious side effects [7].

FCs are contained in citrus fruits together with polymethoxyflavones (PMFs) and coumarins (Cs), the latter also contained in several spices, especially cinnamon. Cs, Fs and PMFs share a similar chemical structure and are named with the common definition of oxygen heterocyclic compounds (OHCs).

All these chemical classes have been widely investigated for their wide range of biological activities. The health effects related to the consumption of PMFs are known enough to make them interesting nutraceuticals for dietary supplements [8].

Furocoumarins inhibit the calcium channel, reduce platelet aggregation [9], induce apoptosis in human promyelocytic leukaemia [10] and show hepatoprotective effects [11]. Imperatorin has been reported to have numerous potent pharmacological actions, including anti-inflammation [12], anti-bacterial [13], anti-allergic [14], beneficial cardiovascular effects [15–17], neuromodulation [18–20] and is able to induce apoptosis in human promyelocytic leukaemia and HL-60 cells [10]. Effects of angelicin against postmenopausal osteoporosis were demonstrated [21]. Bergapten and imperatorin also share in vitro anti-diabetic effects [22]. The FCs found in grapefruit juice show antioxidation, anti-inflammatory and inhibitory activities on breast cancer cell growth, especially bergaptol and bergapten [23]. Different components of bergamot oil showed in vitro activity against neuroblastoma cell growth [7].

Similarly, coumarins possess anti-inflammatory, anti-mutagenic, anti-tumorigenic and antioxidant properties [24–26], moreover have remarkable activities as antihistamine, spasmolysis, inhibition of insulin-induced lipogenesis, antibacterial [1] and anticancer, by acting as inhibitors of tumorigenesis as demonstrated in vivo [27,28] and in vitro [29]. However, in some cases, they showed hepatic [30,31] and pulmonary toxicity [32].

Indeed, as largely reported, Cs and FCs possess several beneficial properties. However, recent studies are focused on the evaluation of the adverse effects as a consequence of their dietary intake.

Among Cs, coumarin is the most investigated because of the harmful effects consequent to its dietary intake [32]. Currently, the legal provision concerning this field is the *Regulation (EC) No 1334/2008* of the European Parliament and of the Council of 16 December 2008 on flavourings and certain food ingredients with flavouring properties for use in and on foods, which includes only coumarin in Annex III, indicating the maximum level permitted in several foods (bakery ware, cereals and desserts) [33]. Most recently [34], the European Food Safety Authority (EFSA) concluded to maintain the tolerable daily intake (TDI) of 0.1 mg coumarin/kg [35] already established in the 2004 opinion [36]. According to the official regulation, the content of coumarin was already determined in many food matrices [31,37]. Whereas, for the Food and Drug Administration (FDA), "food containing any added coumarin as such or as a constituent of tonka beans or tonka extract is deemed to be adulterated" according to *Title 21—Food and Drugs Chapter I—Food and Drug Administration Department of Health and Human Services*.

No limits are imposed on the content of FCs in food, but as mentioned above, they can seriously affect drug metabolism due to the inhibition of the isoform CYP3A4 of cytochrome P450 [38] and the inhibition of efflux transporters [39]. The increase in bioavailability of many drugs can lead to cardiac disorders and respiratory depression [6]. Moreover, FCs have for a long time been known for their phototoxic reactions [40–42], such as phytophotodermatitis [43], due to the formation of intermediates which react with nucleophilic groups such as DNA, RNA and proteins [44].

The bioavailability of dietary FCs ranges from 54 to 81% [45] and there are many evidences about their distribution in the skin [46,47], meaning that also the oral administration of FCs may lead to skin pathologies after UV exposure.

Already in 1986, the International Agency for Research on Cancer (IARC) added 8-methoxypsoralen (8-MOP) into the group one list, as carcinogenic to humans. Despite the European Medicines Agency [48] and the European Food Safety Authority's [49] suggestions, the European Parliament did not impose any limit of their content in food.

However, taking into account all the evidence cited above, the characterization of Cs and FCs in food and beverages is a topic of great interest. Until now, HPLC-DAD methods were widely applied for the quality control of citrus essential oils [50,51]. However, in derived food or cosmetics products, viz. for target analysis in complex samples, more sensitive and selective methods are needed, mainly based on the use of tandem MS systems that allow to achieve low limits of quantification (LOQ) [1,2,20,52–55].

The most comprehensive study about the content of FCs in food was carried out by Melough and co-workers, by using an UPLC-MS/MS system [54] to determine seven major FCs in 19 different foods, some of them already investigated [3].

The aim of the present work is to achieve the comprehensive determination of a total of 35 OHCs, among Cs, Fs and PMFs in different beverages, by using a novel HPLC-MS/MS method with an embebbed linear retention index (LRI) approach.

The LRI system is routinely used in gas chromatography since it has been introduced for the first time [56]; it was abandoned after a few attempts in LC due to a poor repeatability at inter-laboratory levels [57]. Recently, it was re-proposed in our laboratory for the identification of lipids [58] and OHC compounds [51,52,59] through LC methods combined with different detection systems, being confident about the highest batch-to-batch reproducibility of LC columns and instrumentations achieved in the last years. Indeed, the built LRI databases proved to be stable in different instrumental setups [51,52,55]. Within this context, the present research was aimed to characterize several citrus beverages and a marmalade sampled by HPLC-MS/MS; thus, combining the use of an MS/MS library and the LRI database for a reliable identification, even at trace level, below the limit of identification of the common UV detectors. Quantification was performed in Multiple Reaction Monitoring (MRM) mode by external calibration for all the 35 target OHCs. The obtained data add useful information to evaluate the FC dietary intake and the related risk assessment.

2. Materials and Methods

A total of 41 standards among OHCs and alkyl aryl ketones used as reference homologue series for the building of the LRI database were furnished by Merck Life Science (Merck KGaA, Darmstadt, Germany). The full names are provided in Table 1. Tetrahydrofuran (THF), ethyl acetate and ethanol, all HPLC grade, water and methanol both UHPLC-MS grade were provided by Merck Life Science. OHC solutions were prepared by using ethanol as a solvent, whereas alkyl aryl ketones were dissolved in acetonitrile. All standards and stock solutions were maintained at −18 °C before use.

Table 1. MRM conditions (transition, voltages and collision energies). Q, quantifier ion; q, qualifier ion; CE, collision energy, and inter-sample LRI average (Δ LRI) for all the analytes in comparison with tabulated values. SIM, Single Ion Monitoring.

Compound	Class	MRM Transitions			Tabulated LRI	Inter-Sample LRI Average	Δ LRI
		$(M+H)^+$	Q (CE)	q (CE)			
Target compounds							
Meranzin hydrate	C	279	189 (−17)	261 (−7)	783	781	2
Herniarin	C	177	121 (−21)	77 (−25)	799	801	2
Byakangelicin	FC	317 334.7	231 (−19)	233 (−13) 174.95 (−31)	825	824	1
8-methoxypsoralen	FC	217	202 (−21)	174 (−25)	837	830	3
Psoralen	FC	187	131 (−21)	77 (−40)	842	844	2
Angelicin	FC	187	131 (−25)	77 (−35)	853	854	1
Oxypeucedanin hydrate	FC	305	203 (−20)	147 (−32)	864	861	3
Citropten	C	207	192 (−20)	163 (−15)	874	874	0
Isopimpinellin	FC	247	217 (−25)	232 (−18)	884	886	2
Meranzin	C	261	189 (−15)	131 (−29)	886	885	1
Isomeranzin	C	261	189 (−17)	131 (−30)	900	900	0
Heraclenin	FC	287	203 (−17)	147 (−33)	904	907	3

Table 1. *Cont.*

Compound	Class	MRM Transitions			Tabulated LRI	Inter-Sample LRI Average	Δ LRI
		(M + H)+	Q (CE)	q (CE)			
Bergapten	FC	217	202 (−19)	174 (−25)	911	912	1
Sinensetin	PMF	373	343 (−30)	312 (−21)	937	938	1
Isobergapten	FC	217	202 (−21)	174 (−26)	940	942	2
Byakangelicol	FC	317	218 (−29)	175 (−25)	950	954	4
Oxypeucedanin	C	287	203 (−18)	59 (−38)	974	971	3
Nobiletin	PMF	403	373 (−34)	327 (−31)	1016	1017	1
Tetra-O-methylscutellarein	PMF	343	313 (−30)	282 (−25)	1027	1029	2
Trioxsalen	FC	229	142 (−25)	173 (−22)	1076	1075	1
Imperatorin	FC	271	203 (−15)	147 (−31)	1082	1079	3
Tangeretin	PMF	373	343 (−30)	211 (−34)	1087	1085	2
Epoxyaurapten	C	315	163 (−16)	107 (−25)	1096	1096	0
5-O-demethylnobiletin	PMF	389	359 (−30)	341 (−27)	1100	1102	2
Phellopterin	FC	301	233 (−14)	218 (−30)	1104	1105	1
Cnidilin	FC	301	233 (−15)	218 (−28)	1129	1126	3
Gardenin A	PMF	419	389 (−32)	371 (−28)	1130	1134	4
Isoimperatorin	FC	271	203 (−15)	147 (−31)	1159	1160	1
Epoxybergamottin	FC	355	203 (−18)	215 (−19)	1166	1162	4
Gardenin B	PMF	359	329 (−29)	311 (−25)	1172	1170	2
Cnidicin	FC	355	219 (−16)	173(−32)	1303	1305	2
8-geranyloxypsoralen	FC	339	203 (−25)	95 (−25)	1316	1316	0
Aurapten	C	299	163 (−15)	107 (−40)	1336	1339	3
Bergamottin	FC	339	203 (−14)	147 (−35)	1355	1358	3
5-geranyloxy-7-methoxycoumarin	C	329	193 (−20)	149 (−25)	1364	1366	2
Homologous series			SIM (M − H)+				
Acetophenone	AAK		121		800		
Propiophenone	AAK		135		900		
Butyrophenone	AAK		149		1000		
Valerophenone	AAK		163		1100		
Hexanophenone	AAK		177		1200		
Heptanophenone	AAK		191		1300		

2.1. Samples and Sample Preparation

Three bergamot alcoholic beverages, a lemon and a bergamot commercial juices, an Earl grey tea, a citrus infusion (composed of different dried fruits, not specified on the label) and a lemon marmalade were bought in a local market.

A home-made *limoncello* was prepared by ethanol maceration of the lemon peels for around 30 days, then the extract was added to water and sugar.

Juices were centrifuged at 3000 rpm for 5 min (bench top centrifuge, Neya 16R) to remove the pulp, then the supernatant was injected without further pre-treatment. All other samples were subjected to liquid–liquid extraction with ethyl acetate. In particular, 10 mL of alcoholic beverages and 150 mL of infusions were extracted with 10 mL and 50 mL of solvent, respectively. The extraction was carried out by manual shaking in a separatory funnel, the aqueous or hydro-alcoholic phase was iteratively extracted three times with the same amount of ethyl acetate. The three ethyl acetate phases were pooled and dried through a rotary evaporator (Envì EZ-2, Genevac, Ipswich, UK), applying a pre-set method called "low boiling point".

Infusions were prepared by adding 150 mL of boiling water to one sachet, containing two grams of dried material. The infusion time was 5 min.

In the case of marmalade, 10 g of the product were weighed directly in a falcon, added to 10 mL of ethyl acetate and sonicated for 15 min (frequency, kHz 80); the mixture was filtered and the residue was extracted again. The procedure was repeated one more time.

The supernatans were pooled and dried through the Envì EZ-2 rotary evaporator, applying the "low boiling point" method.

All the residues were dissolved in 1 mL of ethanol before the HPLC-MS/MS analysis and three replicates were performed.

2.2. Instrumental and Analytical Conditions

The instrumental setup and the employed method were the same previously developed and validated [52].

The instrument was a Nexera X2 system coupled with a triple quadrupole mass spectrometer LCMS-8060 (Shimadzu, Duisburg, Germany) via an APCI interface set in positive ionization mode. The chromatographic system was equipped with two LC-30AD pumps, a SIL-30AC autosampler, a DGU-20A$_{5R}$ degassing unit and a CTO-20AC oven. The separation was achieved by using an Ascentis Express C18 column (50 × 4.6 mm, 2.7 µm) provided by Merck. The mobile phase was A) water/methanol/THF (85:10:5, *v:v:v*) and B) methanol/THF (95:5, *v:v*). The chromatographic run was carried out at a flow rate of 2 mL min^{-1} and a temperature of 40 °C in gradient mode according to the following program: 0–4.5 min, 15–28% B; 4.5–7.0 min, 28–60% B; 7.0–11.0 min, 60–85% B, hold for 3 min. The injection volume was 2 µL.

The MS system operated in both full scan and MRM acquisition mode to ensure both untargeted and targeted analyses, in full scan and MRM mode, respectively. MS parameters were as follows: interface temperature was set at 450 °C; desolvation line (DL) and heat block temperatures were both 300 °C; nebulizing and drying gas flow were 3 and 15 L/min, respectively; the pressure of the CID gas was 270 kPa. The mass spectral range was 150–450 m/z for the untargeted analysis, while the targeted analysis was carried out in MRM mode through a synchronized method, which set specific acquisition windows according to the retention time of each target, and by applying a dwell time of 20.0 ms. In this way it was possible to obtain 10 scan per peak, as required for a correct quantification. Instead, the homologous series was analysed in Single Ion Monitoring positive mode (SIM+). The m/z monitored for the alkyl aryl ketones was: m/z 121, m/z 135, m/z 149, m/z 163, m/z 177, m/z 191 for acetophenone, propiophenone, butyrophenone, valerophenone, hexanophenone and heptanophenone, respectively. Target analytes (Cs, FCs and PMFs) were detected in MRM acquisition mode.

Taking into account the sensitivity of the MRM detection, the carry-over phenomenon was considered by injecting a reagent blank, represented by 2 µL of pure ethanol between two consecutive sample analyses, in order to evaluate the efficacy of the washing gradient step.

2.3. Qualitative and Quantitative Analysis

Peak identification was carried out by complementarily using the MS/MS library and the LRI database, as reported in the previous work [52]. The homologous series of alkyl aril ketones was injected before the samples and used for the calculation of the LRI values by means of the following equation, according to the Van den Dool and Kratz theory and as previously reported [51,52,59]:

$$LRI = 100 \times \left[z + \frac{t_{Ri} - t_{Rz}}{t_{R(z+1)} - t_{Rz}} \right] \quad (1)$$

where z is the carbon number of the alkyl aryl ketone eluted before the analyte, t_{Ri}, t_{Rz} and $t_{R(z+1)}$ are the retention times of the analyte and the alkyl aryl ketones eluted before and after the analyte.

A new software, namely ChromLinear (Chromaleont, Messina, Italy) version 1.0, developed ad hoc for LRI database handling in LC [60], was employed for a fast and automatic peak identification. The software is able to perform a dual-filter search, thus, excluding from the list of compounds with a high spectral similarity (minimum direct and reverse match were both set at a value of 800; they are calculated based on the NIST MS search algorithm), those falling out the window of ±4 LRI units.

The quantification was based on the creation of calibration curves built in MRM mode for all standards [52].

3. Results

Figures 1 and 2 show the qualitative profiles of two beverage samples, the home-made *limoncello* and the bergamot juice, respectively. The enlargement of the figure shows the components contained in a low amount, which were quantitatively determined through the sensitivity of this technique.

The full scan acquisition mode was used to verify potential co-elutions with the target compounds during the validation process. Table 1 reports the MRM transitions applied for the MS acquisition and the inter-sample LRI average for all the analytes in comparison with the tabulated values. The LRI system was applied for the unambiguous identification of target compounds as a library filter. A maximum difference of four units was obtained for all the compounds, thus, pointing out a satisfactory reproducibility of the LRI approach. For the first time, a novel software, *ChromLinear*, was applied to MS/MS analyses and the identification was automatically performed by match with MS/MS libraries in combination with the LRI database. As an example, Figure 3A shows the spectral similarity search results obtained for peak three (bergapten), pointing out that a list of three candidates was obtained, due to the fact that three isomers are characterized by quite identical fragmentation (same precursor and product ions, both quantifier and qualifier). The application of the additional LRI filter allows the univocal identification due to the automatic exclusion of two isomers characterized by a totally different retention index (Figure 3B). It is noteworthy that the use of the simple retention time could not represent a solution since at inter-laboratory levels or in different periods the retention time could be quite different, while it was already demonstrated that the LRI remains constant because of the normalization effect of the reference homologue series.

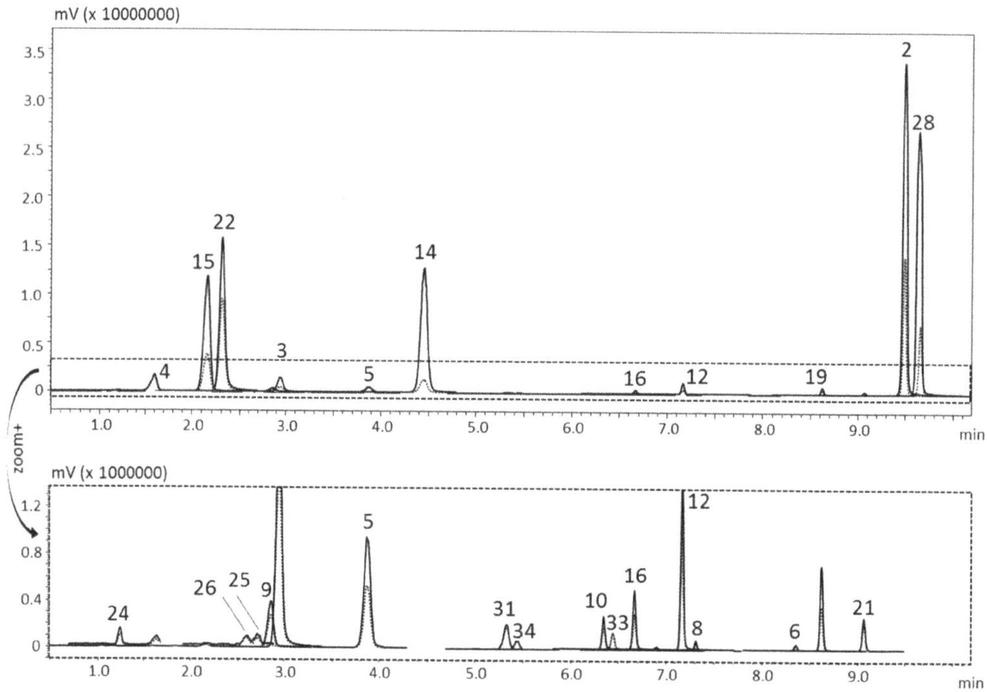

Figure 1. HPLC-MS/MS (MRM acquisition mode) chromatogram of the home-made *limoncello*. Quantifier ion, continuous line; qualifier ion, dotted line. For peak identification see Table 2.

Table 2. Amount of OHCs (mg/L), with the corresponding standard deviation, in the samples analysed. Standard deviation less than 0.001 was not reported. LOQ, limit of quantification.

Compound	Class	Alcoholic Beverages				Infusions		Commercial Juices		Lemon Marmalade
		Home-Made *Limoncello*	Bergamot Beverage A	Bergamot Beverage B	Bergamot Beverage C	Earl grey	Citrus Fruits	Bergamot	Lemon	
Angelicin (1)	FC									
Bergamottin (2)	FC	3.93 ± 0.67	3.50 ± 0.25	2.04 ± 0.02	1.28 ± 0.07	0.001	0.001	19.41 ± 0.91	0.32 ± 0.002	1.16 ± 0.065
Bergapten (3)	FC	0.11 ± 0.007	9.18 ± 0.64	2.03 ± 0.02	0.31 ± 0.01	0.01	0.008	9.74 ± 0.26	<LOQ	<LOQ
Byakangelicin (4)	FC	3.91 ± 0.3		0.11 ± 0.001	0.12 ± 0.004	0.0004	0.01		0.36 ± 0.02	16.3 ± 2.35
Byakangelicol (5)	FC	1.29 ± 0.3			0.06 ± 0.002		0.0003			
Cnidicin (6)	FC	0.14 ± 0.01			0.01 ± 0.001		<LOQ		0.04	
Cnidilin (7)	FC	0.01 ± 0.002			<LOQ	<LOQ	<LOQ		<LOQ	
Epoxybergamottin (8)	FC	0.40 ± 0.003	1.26 ± 0.003		0.003	<LOQ	0.0001	0.03 ± 0.001		
Heraclenin (9)	FC	0.79 ± 0.14			0.01	0.0003				
Imperatorin (10)	FC	0.01 ± 0.002			0.002				0.01 ± 0.004	
Isobergapten (11)	FC									
Isoimperatorin (12)	FC	0.16 ± 0.007			0.01	<LOQ	0.00005	0.01 ± 0.001	0.01	<LOQ
Isopimpinellin (13)	FC	0.003	0.001	0.002	0.002	0.0001	0.00002	0.01	<LOQ	<LOQ
Oxypeucedanin (14)	FC	2.25 ± 1.42		<LOQ	0.11 ± 0.004					
Oxypeucedanin hydrate (15)	FC	3.69 ± 0.22	0.06 ± 0.001	0.07 ± 0	0.10 ± 0.004	0.0004	0.008	0.03	0.22 ± 0.003	12.4 ± 3.01
Phellopterin (16)	FC	0.80 ± 0.008		<LOQ	0.02 ± 0.001		<LOQ		<LOQ	
Psoralen (17)	FC		0.01 ± 0.001	<LOQ	<LOQ	0.0003	<LOQ	0.06 ± 0.001		<LOQ
Trioxsalen (18)	FC									
8-geranyloxypsoralen (19)	FC	6.34 ± 0.39		<LOQ	0.19 ± 0.002		0.0003		0.13 ± 0.01	<LOQ
8-methoxypsoralen (20)	FC			<LOQ	<LOQ	<LOQ				<LOQ
Tot of FC		23.9 ± 3.47	14.1 ± 0.89	4.26 ± 0.04	2.22 ± 0.1	0.01	0.03	29.3 ± 1.17	1.08 ± 0.04	29.86 ± 5.43
Aurapten (21)	C	0.02 ± 0.002	0.001	<LOQ	0.003 ± 0		<LOQ	0.03 ± 0.001		
Citropten (22)	C	3.98 ± 0.39	0.94 ± 0.11	0.58 ± 0.003	0.51 ± 0.02	0.008	0.01	0.78 ± 0.01	0.18 ± 0.01	13.5 ± 2.1
Epoxyaurapten (23)	C									
Herniarin (24)	C	0.01 ± 0.002	0.01 ± 0.001	0.03	0.02	0.003	0.0002	0.01	<LOQ	<LOQ
Isomeranzin (25)	C	<LOQ		0.42	<LOQ		0.03			
Meranzin (26)	C	<LOQ		<LOQ	<LOQ		0.0004			
Meranzin hydrate (27)	C			0.59 ± 0.03		0.001	0.03			
5-geranyloxy-7-methoxycoumarin (28)	C	2.88 ± 0.24	0.19 ± 0.02	0.16 ± 0.003	0.28 ± 0.007	0.0001	0.0004	0.54 ± 0.01	0.27 ± 0.02	0.54 ± 0.06
Tot of C		6.89 ± 0.63	1.14 ± 0.12	1.77 ± 0.03	0.81 ± 0.02	0.01	0.08	1.36 ± 0.02	0.46 ± 0.03	14.04 ± 2.16
Gardenin A (29)	PMF			<LOQ	0.02		0.003	0.07		
Gardenin B (30)	PMF			<LOQ	<LOQ	<LOQ	<LOQ			
Nobiletin (31)	PMF	<LOQ	0.05 ± 0.006	0.92 ± 0.01	0.3 ± 0.02	0.001	0.4 ± 0.01	0.07 ± 0.001	0.02	0.45 ± 0.024
Sinensetin (32)	PMF		0.97 ± 0.01	0.14 ± 0.002	0.02 ± 0.002	0.0002	0.05	0.33 ± 0.01		0.17 ± 0.023
Tangeretin (33)	PMF	0.063	0.07 ± 0.001	0.13 ± 0.001	0.04	0.0003	0.02	0.04 ± 0.001	0.03	0.31 ± 0.001
Tetra-O-methylscutellarein (34)	PMF	<LOQ	0.11 ± 0.001	0.11	0.04 ± 0.002	0.0004	0.02	0.06		0.55 ± 0.004
5-O-demethylnobiletin (35)	PMF			0.13	0.02	0.0004	0.01	0.09 ± 0.003	0.05	
Tot of PMF		0.06	1.20 ± 0.02	1.43 ± 0.01	0.44 ± 0.03	0.002	0.48 ± 0.01	0.65 ± 0.01	0.10	1.48 ± 0.052
TOT		30.8 ± 4.13	16.3 ± 1.04	7.46 ± 0.09	3.47 ± 0.15	0.02	0.59 ± 0.01	31.3 ± 1.20	1.64 ± 0.07	42.7 ± 7.64

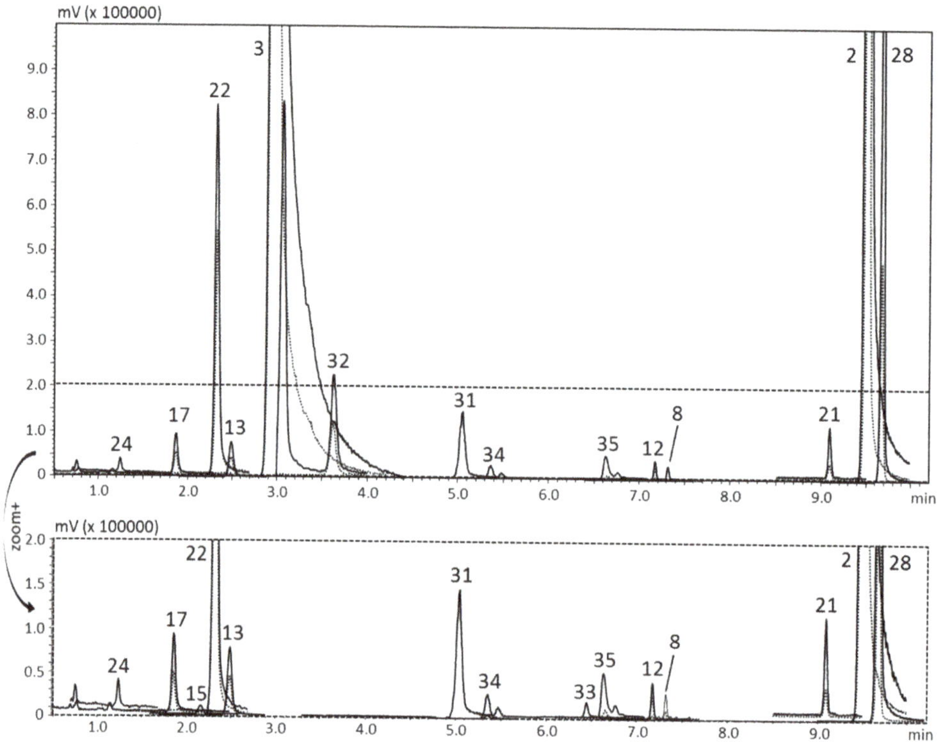

Figure 2. HPLC-MS/MS (MRM acquisition mode) chromatogram of the bergamot juice. Quantifier ion, continuous line; qualifier ion, dotted line. For peak identification see Table 2.

The validated method ensured a wide linearity range thanks to the application of the weighing factor, in order to correct the data heteroscedasticity. LOQ were in the range of 0.0003 and 0.009 mg L^{-1} for most compounds, except for a few exceptions (e.g., nobiletin, 0.0740 mg L^{-1}). Recovery was satisfactory at four levels of concentration tested. More details about the validation parameters are available in the previous study by Arigò and co-workers [52].

Table 2 reports the amounts of OHCs contained in all the samples analysed.

In agreement with the study carried out by Dugo and co-workers [61], bergamot juice resulted in the richest sample with a total amount of 31 mg/L of OHCs, mostly represented by bergamottin and bergapten, 19 and 10 mg/L, respectively. Whereas the lemon juice showed a total of 1.64 mg/L among FCs, Cs and PMFs.

As for alcoholic drinks, the maceration of lemon flavedo in ethanol, as the traditional procedure to prepare *limoncello* in southern Italy, seems to extract a notable amount of OHCs. In particular, the most representative are FCs with a total of 24 mg/L, mainly 8-geranyloxypsoralen, bergamottin, biakangelicin, and oxypeucedanin hydrate (6.3, 3.9, 3.9 and 3.7 mg/L, respectively). Citropten (4.0 mg/L) and 5-geranyloxy-7-methoxycoumarin (2.9 mg/L) constitute almost the overall total for Cs, equal to 6.9 mg/L. The amount of biakangelicin, oxypeucedanin hydrate and citropten quantified in this sample of *limoncello* are very similar to those reported by Dugo [61] for another home-made *limoncello* samples.

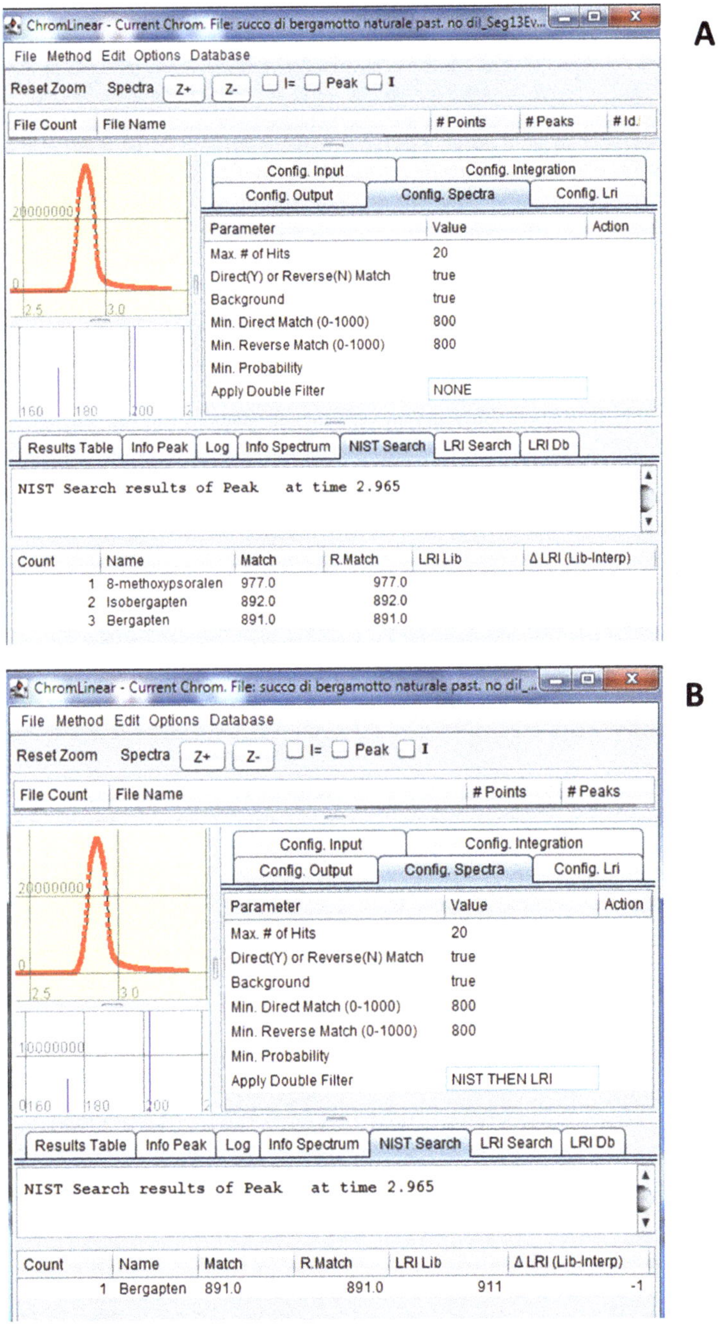

Figure 3. (**A**) Spectral similarity search results and (**B**) dual-filter search results for peak 3 (bergapten).

Three different commercial bergamot alcoholic beverages were analysed in this study. The samples were purchased in local markets sited in Calabria, the only region where *Citrus bergamia* Risso plants are grown. Commercial *bergamino* samples gave very different results, pointing out the differences due to the production process or to the use of different ingredients. The samples showed a very different OHC composition with a total of 16.3, 7.5 and 3.5 mg/L for bergamot beverage A, B and C, respectively. In all cases, much of the total was represented by FCs, 14.0, 4.0 and 2.0 mg/L for sample A, B and C, respectively. Bergamottin and bergapten were the most abundant FCs in all products. Citropten was the major coumarin in all the samples (0.94, 0.58, 0.51 mg/L for sample A, B and C respectively), followed by 5-geranyloxy-7-methoxycoumarin (0.19, 0.16, 0.28 mg/L for sample A, B and C, respectively). Sample B is characterized by the presence of meranzin hydrate and isomeranzin in quite high amounts, 0.59 mg/L and 0.42 mg/L, respectively; both compounds were not detected in the other bergamot liqueurs.

PMFs were different both in qualitative and quantitative composition and the total amount was very low for all samples. Nobiletin was the most abundant PMF in sample B (0.92 mg/L) whereas sinensetin was the most present (0.97 mg/L) in sample A, followed by tetra-O-methylscutellarein (0.11 mg/L). Likely, OHCs extracted from the flavedo of citrus through a maceration process should reflect the OHC profiles of the corresponding peels than of the essential oils. In this regard, sample A was that one which better corresponds to the OHC composition of the bergamot cold-pressed essential oil [50,52].

A total of OHCs equal to 0.02 mg/L was quantified in the tea, whereas around 0.6 mg/L was found in the citrus peel infusion. Among the classes of OHC in the infusion, the main was PMFs.

Table 2 also reports the quantitative data for the lemon marmalade, while Figure 4 shows its MRM chromatographic profile. Byakangelicin is the most abundant compound (16.3 mg/kg) and together with oxypeucedanin hydrate (12.4 mg/kg) represents the main constituents of the FC class, resulting in a total of 29.9 mg/kg. Citropten is the main coumarin (13.5 mg/kg) and some PMFs are contained, but in a low amount.

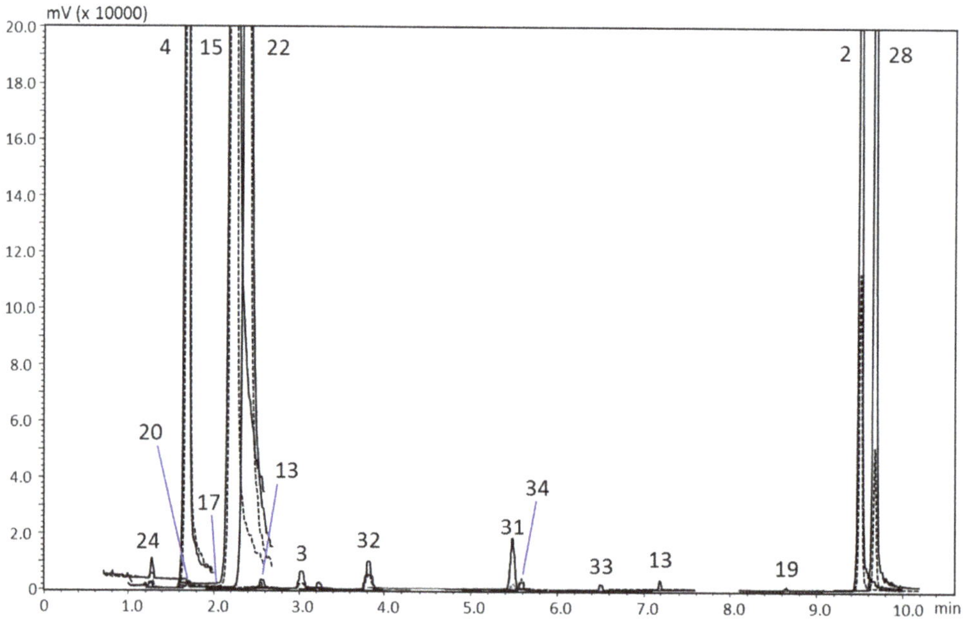

Figure 4. HPLC-MS/MS (MRM acquisition mode) chromatogram of the lemon marmalade. Quantifier ion, continuous line; qualifier ion, dotted line. For peak identification see Table 2.

Recently, the same approach was successfully applied to characterize OHCs in flavoured citrus beer, also in order to prevent food fraud by comparing the volatile profile of the samples with the not volatile composition [52].

4. Discussion

Several scientific articles conclude that there is a lack concerning the in-depth characterization of FCs in foodstuff [40,41,54,62,63].

The object of this research was based on the analysis of several drinks, which can cause the ingestion of considerable amounts of FCs.

The analytical strategy applied in this study was the HPLC-MS/MS method previously developed and successfully applied for the quality control of FCs in beers and cosmetics [52,55]. The sensitivity of the MRM acquisition mode guarantees the correct quantification of OHCs at a very low concentration level. Moreover, the main novelty of this approach arises in the use of the LRI system in combination with MS libraries for the rapid and unambiguous identification of the targets.

We focused the sampling mainly on citrus drinks because OHCs are contained especially in the fruit of this botanical genus. The OHC profile is different depending on the species and the part of the fruit considered, i.e., peel or pulp.

The bergamot fruit is used mainly for the production of essential oil, then the rest of the fruit is generally considered waste [64]. However, in the last years, the production of bergamot beverages is increasing and the consequent dietary intake of FCs should be carefully evaluated.

In order to characterize bergamot fruit and its by-products, in 2016 Russo and coworkers [64] reported a total amount of OHCs in bergamot juice equal to 60 mg/kg, where bergapten and bergamottin were 22 and 37 mg/kg, respectively. The total amount of OHCs in peels was equal to 474 mg/kg, mainly represented by bergapten and bergamottin 219 and 217 mg/kg, respectively.

Flavoured bergamot drinks can be produced by using the essential oil, the juice or the peel of the fruit as ingredients of the preparation, leading to a different content of OHCs. For instance, the maceration of bergamot peels with ethanol, as a traditional method to prepare *bergamino* liqueur, leads to a high content of FCs, as a consequence of the bergapten and bergamottin amounts in the flavedo. Often, commercial alcoholic beverages are not produced by maceration of the bergamot peels, but through the addition of distilled or cold-pressed bergamot essential oils, in this case the amount of OHCs is directly related to the type and the amount of the essential oil used. When distilled essential oils are used, the FC content should be negligible, while for cold-pressed essential oil the content of FCs could grow up to thousands of mg/L.

The same consideration can be applied to other types of beverages flavoured with different species of citrus essential oils (lemon, lime, grapefruit), with the exception of beverages flavoured with orange and mandarin essential oils [8,65] that do not contain FCs.

According to the last toxicological evaluation [38–42], the quality control of these compounds in food represents an emerging subject and the development of an analytical method suitable for this purpose is a challenge. The information derived by the characterization of several types of beverages provides a starting point to discuss the dietary intake of FCs and below which value it can be considered safe.

Versari and co-workers performed the characterization of twelve *Limoncello* liqueurs provided from commercial sources; only six OHCs were detected and quantified [66]. The most recent evaluation of the FC amount contained in a variety of citrus drinks was realized by Gorgus and co-workers [67]. The dietary intake of FCs caused by the ingestion of grapefruit juice was estimated to be higher compared to those caused by the ingestion of other citrus-containing beverages. However, only a few compounds were investigated in this study [67].

Our group already carried out research about the determination of OHCs in citrus products by HPLC with UV detection [61]. In that study, different citrus beverages were analysed, but the results obtained were strictly related to the low sensitivity of the detector employed. In particular, 27 OHCs were investigated in a commercial Earl Grey tea, two laboratory-made juices (lemon and bergamot), a home-made *limoncello* and two commercial *bergamino* liquors. *Limoncello* resulted the qualitatively richest, whereas the bergamot juice was surprisingly the most abundant from the quantitative point of view. This depends on the ingredients used for the commercial drinks; in fact, despite the traditional preparation being based on the maceration of flavedo with alcohol which should correspond to the high extraction of FCs, the industrial products may be prepared by using essential oils or juices, explaining the lower amounts of FCs detected in *limoncello* compared to the bergamot juice.

The variables responsible for the differences in the OHC quali-quantitative profile of the analysed samples are numerous. The ingredients of the preparation play the key role on the target composition in the samples. In facts, whereas home-made *limoncello* and *bergamino* are prepared with the traditional recipe which is based on the flavedo maceration with ethanol, the commercial liqueurs may contain essential oils (cold-pressed or distilled), alcoholic extracts and/or juice.

As for infusion products, the quantification of OHCs, especially FCs, has attracted special interest, following a case report published in *The Lancet* journal in 2002 [68]. In this article, the abundant and continuous assumption of earl grey tea was clearly connected with neurologic adverse effects. The reason was attributed to bergapten, which is contained in the bergamot essential oil added to black tea, to confer the characteristic aroma. The author's conclusion was that "even tea may lead to health problems if flavoured and consumed in extraordinarily high quantities".

In this research, an earl grey tea and a citrus fruit infusion were also taken into account. Both samples were extracted prior to the analysis in order to concentrate the target compounds, which are contained in low amounts mainly because of their low water solubility, then the calculated concentrations (above the LOQ) were corrected for the dilution factor. The infusion of dried citrus peels resulted richer, than the earl grey tea, both in FCs and total OHCs. This means that the powder could be composed mainly of dried peels of mandarin and orange fruits. The reason is that often, to avoid the presence of FCs, peels of sweet orange or mandarin fruits are preferred, also for the lower price and especially for their sweet taste compared to other types of citrus fruit.

In this sample of earl grey tea, the amounts of bergamottin and bergapten were very low. The ingredients of the product, in particular the essence used to flavour the black tea, were not specified on the label, as it often happens, but we can suppose that the producers used cold-pressed essential oils, which justifies the presence of OHCs.

To summarize, among the beverages analysed, bergamot juice was the richest in FCs, followed by the home-made *limoncello*, with 29 and 24 mg/L, respectively.

Moreover, for the absolute amounts of the targets in the samples, it is important to consider the type of drink, which is correlated with the volume generally ingested.

For instance, the consumption of 20 mL of *limoncello* makes 0.6 mg of FCs, but drinking 200 mL of juice causes the makes 4.8 mg of FCs.

A dietary intake of around 20 g of marmalade causes the ingestion of 0.6 mg of FCs.

5. Conclusions

Due to the lack of detailed data about the content of FCs and Cs in food, the opinions regarding the maximum admitted level are still varied and contrasting, consequently an official regulation is still missing.

This HPLC-MS/MS method, characterized by the application of the LRI system combined with the MS libraries, was successfully applied to determine three different classes of OHCs in different citrus beverages and marmalade, representing a significant step forward with respect to previous works which are mainly focused on a limited number of compounds. The highest sensitivity of the MRM method allowed for the detection of a major number of compounds, thus, resulting in a more accurate evaluation of the total amount of FCs and Cs. Their quantification represents the main purpose of the present research, aimed to add data that can be used to estimate their dietary intake.

The method will be applied to other food matrices and new analytical technologies, as supercritical fluid chromatography coupled to triple quadrupole mass spectrometry, will be applied and compared with the present method.

Author Contributions: Investigation, A.A. and E.T.; data curation, A.A.; writing—original draft preparation, A.A., M.R. and F.R.; writing—review and editing, A.A. and F.R.; supervision, P.D. and L.M. All authors have read and agreed to the published version of the manuscript.

Funding: This research received no external funding.

Institutional Review Board Statement: Not applicable.

Informed Consent Statement: Not applicable.

Data Availability Statement: Not applicable.

Acknowledgments: The authors gratefully acknowledge the Shimadzu Corporation and Merck Life Science (Merck KGaA, Darmstadt, Germany) for their continuous support.

Conflicts of Interest: The authors declare no conflict of interest.

References

1. Zheng, X.; Zhang, X.; Sheng, X.; Yuan, Z.; Yang, W.; Wang, Q.; Zhang, L. Simultaneous characterization and quantitation of 11 coumarins in Radix Angelicae Dahuricae by high performance liquid chromatography with electrospray tandem mass spectrometry. *J. Pharm. Biomed. Anal.* **2010**, *51*, 599–605. [CrossRef]
2. Kang, J.; Zhou, L.; Sun, J.; Han, J.; Guo, D.A. Chromatographic fingerprint analysis and characterization of furocoumarins in the roots of Angelica Dahurica by HPLC/DAD/ESI-MSn technique. *J. Pharm. Biomed. Anal.* **2008**, *47*, 778–785. [CrossRef] [PubMed]
3. Wagstaff, D.J. Dietary exposure to furocoumarins. *Regul. Toxicol. Pharmacol.* **1991**, *14*, 261–272. [CrossRef]
4. Peroutka, R.; Schulzová, V.; Botek, P.; Hajšlová, J. Analysis of furanocoumarins in vegetables (Apiaceae) and citrus fruits (Rutaceae). *J. Sci. Food Agric.* **2007**, *87*, 2152–2163. [CrossRef]
5. Ramírez-Pelayo, C.; Martínez-Quiñones, J.; Gil, J.; Durango, D. Coumarins from the peel of citrus grown in Colombia: Composition, elicitation and antifungal activity. *Heliyon* **2019**, *5*, e01937. [CrossRef]
6. Dugrand-Judek, A.; Olry, A.; Hehn, A.; Costantino, G.; Ollitrault, P.; Froelicher, Y.; Bourgaud, F. The distribution of coumarins and furanocoumarins in Citrus species closely matches Citrus phylogeny and reflects the organization of biosynthetic pathways. *PLoS ONE* **2015**, *10*, e0142757. [CrossRef] [PubMed]
7. Hung, W.L.; Suh, J.H.; Wang, Y. Chemistry and health effects of furanocoumarins in grapefruit. *J. Food Drug Anal.* **2017**, *25*, 71–83. [CrossRef]
8. Russo, M.; Rigano, F.; Arigò, A.; Sciarrone, D.; Calabrò, M.L.; Farnetti, S.; Dugo, P.; Mondello, L. Rapid isolation, reliable characterization, and water solubility improvement of polymethoxyflavones from cold-pressed mandarin essential oil. *J. Sep. Sci.* **2016**, *39*, 2018–2027. [CrossRef]
9. Chen, I.S.; Chang, C.T.; Sheen, W.S.; Teng, C.M.; Tsai, I.L.; Duh, C.Y.; Ko, F.N. Coumarins and antiplatelet aggregation constituents from formosan *Peucedanum japonicum*. *Phytochemistry* **1996**, *41*, 525–530. [CrossRef]
10. Pae, H.O.; Oh, H.; Yun, Y.G.; Oh, G.S.; Jang, S.I.; Hwang, K.M.; Kwon, T.O.; Lee, H.S.; Chung, H.T. Imperatorin, a furanocoumarin from *Angelica dahurica* (Umbelliferae), induces cytochrome c-dependent apoptosis in human promyelocytic leukaemia, HL-60 cells. *Pharmacol. Toxicol.* **2002**, *91*, 40–48. [CrossRef]
11. Najda, A.B.; Dyduch, J.; Sachadyn-Król, M.; Najda, A.; Świca, K.; Kapłan, M.; Papliński, R.; Klimek, K. Identification and profile of furanocoumarins from the ribbed celery (*Apium Graveolens* L. Var. 2 *Dulce* Mill./Pers.). *Food Sci. Technol. Res.* **2015**, *21*, 67–75. [CrossRef]
12. Huang, H. Plant diversity and conservation in China: Planning a strategic bioresource for a sustainable future. *Bot. J. Linn. Soc.* **2011**, *166*, 282–300. [CrossRef]
13. Raja, S.B.; Murali, M.R.; Roopa, K.; Devaraj, S.N. Imperatorin a furocoumarin inhibits periplasmic Cu-Zn SOD of Shigella dysenteriae their by modulates its resistance towards phagocytosis during host pathogen interaction. *Biomed. Pharmacother.* **2011**, *65*, 560–568. [CrossRef]
14. Oh, H.A.; Kim, H.M.; Jeong, H.J. Distinct effects of imperatorin on allergic rhinitis: Imperatorin inhibits caspase-1 activity in vivo and in vitro. *J. Pharmacol. Exp. Ther.* **2011**, *339*, 72–81. [CrossRef]
15. Bertin, R.; Chen, Z.; Martínez-Vázquez, M.; García-Argaéz, A.; Froldi, G. Vasodilation and radical-scavenging activity of imperatorin and selected coumarinic and flavonoid compounds from genus Casimiroa. *Phytomedicine* **2014**, *21*, 586–594. [CrossRef] [PubMed]

16. He, J.Y.; Zhang, W.; He, L.C.; Cao, Y.X. Imperatorin induces vasodilatation possibly via inhibiting voltage dependent calcium channel and receptor-mediated Ca2+ influx and release. *Eur. J. Pharmacol.* **2007**, *573*, 170–175. [CrossRef] [PubMed]
17. Nie, H.; Meng, L.Z.; Zhou, J.Y.; Fan, X.F.; Luo, Y.; Zhang, G.W. Imperatorin is responsible for the vasodilatation activity of Angelica dahurica var. formosana regulated by nitric oxide in an endothelium-dependent manner. *Chin. J. Integr. Med.* **2009**, *15*, 442–447. [CrossRef]
18. Luszczki, J.J.; Glowniak, K.; Czuczwar, S.J. Time-course and dose-response relationships of imperatorin in the mouse maximal electroshock seizure threshold model. *Neurosci. Res.* **2007**, *59*, 18–22. [CrossRef]
19. Wu, K.C.; Chen, Y.H.; Cheng, K.S.; Kuo, Y.H.; Yang, C.T.; Wong, K.L.; Tu, Y.K.; Chan, P.; Leung, Y.M. Suppression of voltage-gated Na+ channels and neuronal excitability by imperatorin. *Eur. J. Pharmacol.* **2013**, *721*, 49–55. [CrossRef]
20. Ngo, L.; Tran, P.; Ham, S.H.; Cho, J.H.; Cho, H.Y.; Lee, Y.B. Simultaneous determination of imperatorin and its metabolite xanthotoxol in rat plasma and urine by LC–MS/MS and its application to pharmacokinetic studies. *J. Chromatogr. B Anal. Technol. Biomed. Life Sci.* **2017**, *1044–1045*, 30–38. [CrossRef]
21. Wang, J.; Wang, G.; Gong, L.; Sun, G.; Shi, B.; Bao, H.; Duan, Y. Isopsoralen regulates PPAR-γ/WNT to inhibit oxidative stress in osteoporosis. *Mol. Med. Rep.* **2018**, *17*, 1125–1131. [CrossRef]
22. Shalaby, N.M.M.; Abd-Alla, H.I.; Aly, H.F.; Albalawy, M.A.; Shaker, K.H.; Bouajila, J. Preliminary in vitro and in vivo evaluation of antidiabetic activity of ducrosia anethifolia boiss. and its linear furanocoumarins. *BioMed Res. Int.* **2014**, *2014*, 480545. [CrossRef]
23. Panno, M.; Giordano, F.; Palma, M.; Bartella, V.; Rago, V.; Maggiolini, M.; Sisci, D.; Lanzino, M.; de Amicis, F.; Ando, S. Evidence that bergapten, independently of its photoactivation, enhances p53 gene expression and induces apoptosis in human breast cancer cells. *Curr. Cancer Drug Targets* **2009**, *2014*, 1–13. [CrossRef]
24. Deng, G.G.; Wei, W.; Yang, X.W.; Zhang, Y.B.; Xu, W.; Gong, N.B.; Lü, Y.; Wang, F.F. New coumarins from the roots of Angelica dahurica var. formosana cv. Chuanbaizhi and their inhibition on NO production in LPS-activated RAW264.7 cells. *Fitoterapia* **2015**, *101*, 194–200. [CrossRef] [PubMed]
25. Liu, D.P.; Luo, Q.; Wang, G.H.; Xu, Y.; Zhang, X.K.; Chen, Q.C.; Chen, H.F. Furocoumarin derivatives from radix angelicae dahuricae and their effects on rxrα transcriptional regulation. *Molecules* **2011**, *16*, 6339–6348. [CrossRef] [PubMed]
26. Bai, Y.; Li, D.; Zhou, T.; Qin, N.; Li, Z.; Yu, Z.; Hua, H. Coumarins from the roots of Angelica dahurica with antioxidant and antiproliferative activities. *J. Funct. Foods* **2016**, *20*, 453–462. [CrossRef]
27. Nair, R.V.; Fisher, E.P.; Safe, S.H.; Cortez, C.; Harvey, R.G.; Digiovanni, J. Novel coumarins as potential anticarcinogenic agents. *Carcinogenesis* **1991**, *12*, 65–69. [CrossRef] [PubMed]
28. Noguchi, M.; Kitagawa, H.; Miyazaki, I.; Mizukami, Y. Influence of esculetin on incidence, proliferation, and cell kinetics of mammary carcinomas induced by 7,12-dimethylbenz[α]anthracene in rats on high- and low-fat diets. *Jpn. J. Cancer Res.* **1993**, *84*, 1010–1014. [CrossRef]
29. Edenharder, R.; Speth, C.; Decker, M.; Kolodziej, H.; Kayser, O.; Platt, K.L. Inhibition of mutagenesis of 2-amino-3-methylimidazo [4,5-*f*]quinoline (IQ) by coumarins and furanocoumarins, chromanones and furanochromanones. *Mutat. Res. Toxicol.* **1995**, *345*, 57–71. [CrossRef]
30. Abraham, K.; Pfister, M.; Wöhrlin, F.; Lampen, A. Relative bioavailability of coumarin from cinnamon and cinnamon-containing foods compared to isolated coumarin: A four-way crossover study in human volunteers. *Mol. Nutr. Food Res.* **2011**, *55*, 644–653. [CrossRef]
31. Krüger, S.; Winheim, L.; Morlock, G.E. Planar chromatographic screening and quantification of coumarin in food, confirmed by mass spectrometry. *Food Chem.* **2018**, *239*, 1182–1191. [CrossRef] [PubMed]
32. Lake, B.G. Coumarin metabolism, toxicity and carcinogenicity: Relevance for human risk assessment. *Food Chem. Toxicol.* **1999**, *37*, 423–453. [CrossRef]
33. European Commission. Regulation (EC) No 1334/2008 on flavourings and certain food ingredients. *Off. J. Eur. Union* **2008**.
34. European Commission. (Text with EEA relevance) 21.6.2017. *Off. J. Eur. Union* **2018**, *2016*, 48–119.
35. European Food Safety Authority (EFSA). Coumarin in flavourings and other food ingredients with flavouring properties—Scientific opinion of the panel on food additives, flavourings, processing aids and materials in contact with food (AFC). *EFSA J.* **2008**, *6*. [CrossRef]
36. European Food Safety Authority (EFSA). Opinion of the Scientific Panel on food additives, flavourings, processing aids and materials in contact with food (AFC) on a request from the Commission related to tertiary-Butylhydroquinone (TBHQ). *EFSA J.* **2004**, *2*, 84. [CrossRef]
37. Sproll, C.; Ruge, W.; Andlauer, C.; Godelmann, R.; Lachenmeier, D.W. HPLC analysis and safety assessment of coumarin in foods. *Food Chem.* **2008**, *109*, 462–469. [CrossRef]
38. Girennavar, B.; Poulose, S.M.; Jayaprakasha, G.K.; Bhat, N.G.; Patil, B.S. Furocoumarins from grapefruit juice and their effect on human CYP 3A4 and CYP 1B1 isoenzymes. *Bioorganic Med. Chem.* **2006**, *14*, 2606–2612. [CrossRef] [PubMed]
39. Strauch, K.; Lutz, U.; Bittner, N.; Lutz, W.K. Dose-response relationship for the pharmacokinetic interaction of grapefruit juice with dextromethorphan investigated by human urinary metabolite profiles. *Food Chem. Toxicol.* **2009**, *47*, 1928–1935. [CrossRef]
40. Wu, S.; Cho, E.; Feskanich, D.; Li, W.Q.; Sun, Q.; Han, J.; Qureshi, A.A. Citrus consumption and risk of basal cell carcinoma and squamous cell carcinoma of the skin. *Carcinogenesis* **2015**, *36*, 1162–1168. [CrossRef]
41. Wu, S.; Han, J.; Feskanich, D.; Cho, E.; Stampfer, M.J.; Willett, W.C.; Qureshi, A.A. Citrus consumption and risk of cutaneous malignant melanoma. *J. Clin. Oncol.* **2015**, *33*, 2500–2508. [CrossRef]

42. Ueng, Y.F.; Chen, C.C.; Huang, Y.L.; Lee, I.J.; Yun, C.H.; Chen, Y.H.; Huang, C.C. Effects of aqueous extract of Ruta graveolens and its ingredients on cytochrome P450, uridine diphosphate (UDP)-glucuronosyltransferase, and reduced nicotinamide adenine dinucleotide (phosphate) (NAD(P)H)-quinone oxidoreductase in mice. *J. Food Drug Anal.* **2015**, *23*, 516–528. [CrossRef] [PubMed]
43. Melough, M.M.; Vance, T.M.; Lee, S.G.; Provatas, A.A.; Perkins, C.; Qureshi, A.; Cho, E.; Chun, O.K. Furocoumarin kinetics in plasma and urine of healthy adults following consumption of grapefruit (*Citrus paradisi* Macf.) and grapefruit juice. *J. Agric. Food Chem.* **2017**, *65*, 3006–3012. [CrossRef]
44. Creeke, P.I.; Ginestar, M.P.C.; Lee, H.A.; Morgan, M.R.A.; Price, K.R.; Rhodes, M.J.C.; Wilkinson, A.P. Development of an immunoassay to measure the absorption of dietary furanocoumarins in humans. *Biochem. Soc. Trans.* **1996**, *24*. [CrossRef]
45. Luo, Q.; Wang, C.P.; Li, J.; Ma, W.F.; Bai, Y.; Ma, L.; Gao, X.M.; Zhang, B.L.; Chang, Y.X. The pharmacokinetics and oral bioavailability studies of columbianetin in rats after oral and intravenous administration. *J. Ethnopharmacol.* **2013**, *150*, 175–180. [CrossRef] [PubMed]
46. Yeo, U.C.; Shin, J.H.; Yang, J.M.; Park, K.B.; Kim, M.M.; Bok, H.S.; Lee, E.S. Psoralen-ultraviolet A-induced erythema: Sensitivity correlates with the concentrations of psoralen in suction blister fluid. *Br. J. Dermatol.* **2000**, *142*, 733–739. [CrossRef] [PubMed]
47. Tegeder, I.; Bräutigam, L.; Podda, M.; Meier, S.; Kaufmann, R.; Geisslinger, G.; Grundmann-Kollmann, M. Time course of 8-methoxypsoralen concentrations in skin and plasma after topical (bath and cream) and oral administration of 8-methoxypsoralen. *Clin. Pharmacol. Ther.* **2002**, *71*, 153–161. [CrossRef]
48. European Medicines Agency. Reflection Paper on the Risks Associated with Furocoumarins Contained in Preparations of *Angelica archangelica* L. *Comm. Herb. Med. Prod.* **2007**, 1–22.
49. Committee, E.S. Scientific Opinion on a Qualified Presumption of Safety (QPS) approach for the safety assessment of botanicals and botanical preparations. *EFSA J.* **2014**, *12*, 1–38. [CrossRef]
50. Russo, M.; Bonaccorsi, I.; Costa, R.; Trozzi, A.; Dugo, P.; Mondello, L. Reduced time HPLC analyses for fast quality control of citrus essential oils. *J. Essent. Oil Res.* **2015**, *27*, 307–315. [CrossRef]
51. Arigò, A.; Rigano, F.; Micalizzi, G.; Dugo, P.; Mondello, L. Oxygen heterocyclic compound screening in Citrus essential oils by linear retention index approach applied to liquid chromatography coupled to photodiode array detector. *Flavour Fragr. J.* **2019**, *34*, 349–364. [CrossRef]
52. Arigò, A.; Dugo, P.; Rigano, F.; Mondello, L. Linear retention index approach applied to liquid chromatography coupled to triple quadrupole mass spectrometry to determine oxygen heterocyclic compounds at trace level in finished cosmetics. *J. Chromatogr. A* **2021**, *1649*, 462183. [CrossRef]
53. Li, L.; Chen, J.; Yang, X.; Zhang, Y.; Yang, D. In vivo metabolite profiles of isoimperatorin and phellopterin in rats analyzed using HPLC coupled with diode array detector and electrospray ionization ion trap time-of-flight mass spectrometry technique. *Biomed. Chromatogr.* **2018**, *32*, e4236. [CrossRef]
54. Melough, M.M.; Lee, S.G.; Cho, E.; Kim, K.; Provatas, A.A.; Perkins, C.; Park, M.K.; Qureshi, A.; Chun, O.K. Identification and quantitation of furocoumarins in popularly consumed foods in the U.S. using QuEChERS extraction coupled with UPLC-MS/MS analysis. *J. Agric. Food Chem.* **2017**, *65*, 5049–5055. [CrossRef] [PubMed]
55. Trovato, E.; Arigò, A.; Vento, F.; Micalizzi, G.; Dugo, P.; Mondello, L. Influence of citrus flavor addition in brewing process: Characterization of the volatile and non-volatile profile to prevent frauds and adulterations. *Separations* **2021**, *8*, 18. [CrossRef]
56. Kováts, E. Gas-chromatographische Charakterisierung organischer Verbindungen. Teil 1: Retentionsindices aliphatischer Halogenide, Alkohole, Aldehyde und Ketone. *Helv. Chim. Acta* **1958**, *41*, 1915–1932. [CrossRef]
57. Rigano, F.; Arigò, A.; Oteri, M.; La Tella, R.; Dugo, P.; Mondello, L. The Retention index approach in liquid chromatography: An historical review and recent advances. *J. Chromatogr. A* **2021**, *1640*, 461963. [CrossRef] [PubMed]
58. Rigano, F.; Oteri, M.; Russo, M.; Dugo, P.; Mondello, L. Proposal of a linear retention index system for improving identification reliability of triacylglycerol profiles in lipid samples by liquid chromatography methods. *Anal. Chem.* **2018**, *90*, 3313–3320. [CrossRef] [PubMed]
59. Rigano, F.; Russo, M.; Arigò, A.; Dugo, P.; Mondello, L. Combining linear retention index and electron ionization mass spectrometry for a reliable identification in nano liquid chromatography. *J. Chromatogr. A* **2020**, *1610*, 460581. [CrossRef] [PubMed]
60. Rigano, F.; Oteri, M.; Micalizzi, G.; Mangraviti, D.; Dugo, P.; Mondello, L. Lipid profile of fish species by liquid chromatography coupled to mass spectrometry and a novel linear retention index database. *J. Sep. Sci.* **2020**, *43*, 1773–1780. [CrossRef]
61. Dugo, P.; Piperno, A.; Romeo, R.; Cambria, M.; Russo, M.; Carnovale, C.; Mondello, L. Determination of oxygen heterocyclic components in citrus products by HPLC with UV detection. *J. Agric. Food Chem.* **2009**, *57*, 6543–6551. [CrossRef]
62. Melough, M.M.; Chun, O.K. Dietary furocoumarins and skin cancer: A review of current biological evidence. *Food Chem. Toxicol.* **2018**, *122*, 163–171. [CrossRef] [PubMed]
63. Prosen, H.; Kočar, D. Different sample preparation methods combined with LC-MS/MS and LC-UV for determination of some furocoumarin compounds in products containing citruses. *Flavour Fragr. J.* **2008**, *23*, 263–271. [CrossRef]
64. Russo, M.; Arigò, A.; Calabrò, M.L.; Farnetti, S.; Mondello, L.; Dugo, P. Bergamot (*Citrus bergamia* Risso) as a source of nutraceuticals: Limonoids and flavonoids. *J. Funct. Foods* **2016**, *20*, 10–19. [CrossRef]
65. Russo, M.; Bonaccorsi, I.; Inferrera, V.; Dugo, P.; Mondello, L. Underestimated sources of flavonoids, limonoids and dietary fiber: Availability in orange's by-products. *J. Funct. Foods* **2015**, *12*, 150–157. [CrossRef]
66. Versari, A.; Natali, N.; Russo, M.T.; Antonelli, A. Analysis of some Italian lemon liquors (Limoncello). *J. Agric. Food Chem.* **2003**, *51*, 4978–4983. [CrossRef]

67. Gorgus, E.; Lohr, C.; Raquet, N.; Guth, S.; Schrenk, D. Limettin and furocoumarins in beverages containing citrus juices or extracts. *Food Chem. Toxicol.* **2010**, *48*, 93–98. [CrossRef]
68. Finsterer, J. Earl Grey tea intoxication. *Lancet* **2002**, *359*, 1484. [CrossRef]

foods

MDPI

Article

Headspace/GC–MS Analysis and Investigation of Antibacterial, Antioxidant and Cytotoxic Activity of Essential Oils and Hydrolates from *Rosmarinus officinalis* L. and *Lavandula angustifolia* Miller

Stefania Garzoli [1,*], Valentina Laghezza Masci [2], Sara Franceschi [3], Antonio Tiezzi [2], Pierluigi Giacomello [1] and Elisa Ovidi [2]

1 Department of Drug Chemistry and Technology, Sapienza University, 00185 Rome, Italy; pierluigi.giacomello@uniroma1.it
2 Department for the Innovation in Biological, Agrofood and Forestal Systems, Tuscia University, 01100 Viterbo, Italy; laghezzamasci@unitus.it (V.L.M.); antoniot@unitus.it (A.T.); eovidi@unitus.it (E.O.)
3 Department of Economics and Statistics, University of Siena, 53100 Siena, Italy; franceschi2@unisi.it
* Correspondence: stefania.garzoli@uniroma1.it

Citation: Garzoli, S.; Laghezza Masci, V.; Franceschi, S.; Tiezzi, A.; Giacomello, P.; Ovidi, E. Headspace/GC–MS Analysis and Investigation of Antibacterial, Antioxidant and Cytotoxic Activity of Essential Oils and Hydrolates from *Rosmarinus officinalis* L. and *Lavandula angustifolia* Miller. *Foods* **2021**, *10*, 1768. https://doi.org/10.3390/foods10081768

Academic Editors: Ian Southwell, Oscar Núñez and Christopher John Smith

Received: 4 July 2021
Accepted: 28 July 2021
Published: 30 July 2021

Publisher's Note: MDPI stays neutral with regard to jurisdictional claims in published maps and institutional affiliations.

Abstract: In this work, essential oils (EOs) and hydrolates (Hys) of *Rosmarinus officinalis* L. and *Lavandula angustifolia* Mill., grown in Tuscany (Italy), were studied to describe their chemical composition and biological activities. The aromatic profile of the EOs liquid phase was carried out by gas chromatography–mass spectrometry (GC–MS), while the volatile composition of vapor phase EOs and Hys was performed by headspace (HS)/GC–MS. The obtained results show that monoterpene hydrocarbons (71.5% and 89.5%) were the main compounds, followed by oxygenated monoterpenes (26.0% and 10.5%) in the liquid and vapor phase of *R. officinalis* EO, respectively. The oxygenated monoterpenes were the main components of *L. angustifolia* EO, reaching 86.9% in the liquid phase and 53.7% in the vapor phase. Regarding Hys, they consisted only of oxygenated monoterpenes, and 1,8-cineole (56.2%) and linalool (42.9%), were the main components of *R. officinalis* and *L. officinalis* Hys, respectively. Their cytotoxicity was investigated on an SHSY5Y neuroblastoma cell line by thiazolyl blue tetrazolium bromide (MTT) test, showing a notable effect of the EOs with a time-independent manner of activity and half maximal effective concentration (EC_{50}) values quite similar for the two plant species (from 0.05% to 0.06% v/v for the three time points evaluated). A measurable activity of Hys was also obtained although with higher EC_{50} values. The antibacterial activity against *Escherichia coli* ATCC® 25922, *Pseudomonas fluorescens* ATCC® 13525, *Acinetobacter bohemicus* DSM 102855 as Gram-negative bacteria and *Kocuria marina* DSM 16420, *Bacillus cereus* ATCC® 10876 as Gram-positive bacteria, was evaluated by the agar disk-diffusion method and the VPT (vapor phase test) to determinate the MIC (minimal inhibitory concentration) and the MBC (minimal bactericidal concentration) values. Both EOs possessed a high activity against all the bacterial strains with MIC values ranging from 0.19% to 3.13% v/v. Unlike EOs, Hys did not show an inhibition of the bacterial growth at the tested concentrations. Furthermore, antioxidant power was measured by 2,2′-azino-bis (3-ethylbenzothiazoline-6-sulfonic acid) diammonium salt-based (ABTS•+) and the 2,2-diphenyl-1-picrylhydrazyl (DPPH) assays, showing a remarkable ability to reduce radicals by both EOs; Hys were slightly less active. The findings highlighted that *R. officinalis* and *L. angustifolia* EOs and Hys have a chemical composition rich in bioactive molecules, which can exert different biological activities.

Keywords: *Lavandula angustifolia*; *Rosmarinus officinalis*; essential oils; hydrolates; chemical analysis; volatile compounds; biological activity

1. Introduction

Since historic times, spices and herbs have been used as food flavors, and their effects on human health are still being investigated to understand the roles of their chemical components [1,2]. *Rosmarinus officinalis* L. (also known as *Salvia rosmarinus* Schleid) and *Lavandula angustifolia* Mill. belong to the Lamiaceae family, which comprises different genera whose biological activities are used in traditional medicines all over the world [3–6]. Rosemary, native to the Mediterranean area, is a shrubby plant up to 1.8 m tall, erect or procumbent with a good aromatic scent due to the glandular hairs that emit volatile essential oils [7]. This plant is widely used in cosmetic preparations to protect from degradation and absorbing UV light, is used as a bactericidal and antifungal agent and furthermore, among others, was exploited in topical applications for wound healing, skin cancer and antimycotic properties [5,7]. Different uses of *R. officinalis* are known and its volatile essential oil (EO) and leaf extracts possess extensively investigated biological properties, such as antioxidant, anti-inflammatory, antiproliferative, anticancer, antiviral, antimicrobial, hepatoprotective, neuroprotective, nephroprotective, antiulcer and many others [8]. *R. officinalis* was investigated for its curative properties against some ailments caused by biochemical, chemical or biological agents as reviewed by Oliviera et al. [9], showing that this plant possesses beneficial effects and may be used to treat health problems. English or "true" lavender, the common names of *Lavandula angustifolia*, is one of the 39 species of the genus Lavandula to which belongs different hybrids [10]. Lavender EO production and quality is regulated by environmental and developmental conditions and temperature and flowering stage determine its chemical composition [11]. *L. angustifolia* EO is used in perfumery and cosmetics and its activity on the central nervous system, as a sedative, anxiolytic and antidepressant, was also evidenced [12–14]. Furthermore, biological activities of EOs from Lavandula genus, such as antifungal, antibacterial, antioxidant and anticancer effects, were reported [15–20]. EO chemical composition is highly complex and it can vary considerably depending on several factors, such as the cultivation area, environmental conditions, morphological characteristics and processing techniques of the plant [21–23]; moreover, the chemical composition influences the way in which EOs exert their antibacterial activity [24]. Terpene and sesquiterpene hydrocarbons, oxygenated or cyclic, are the main classes of compounds present in EOs, followed by aldehydes, ketones, alcohols, acids and esters [25]. In particular, thanks to their interesting physicochemical characteristics, Lamiaceae EOs were employed in the industrial and medical research sectors as natural products [26]. EOs obtained from *L. angustifolia* and *R. officinalis* grown in different countries such as China, Siria, India, Iran, Romania, Canada, Spain, France and others were investigated and linalool, borneol, linalyl acetate and 1,8-cineole as well as camphor, camphene and α-pinene resulted as predominant compounds, although in different proportions according to the vegetative stage and climatic conditions of the origin area [27,28]. There are far fewer studies concerning the volatile chemical composition of hydrolates (Hys), also known as hydrosols. They are aqueous solutions obtained as by-products of distillation [29] containing a certain number of bioactive molecules, although with marked quantitative and qualitative differences compared to EOs [30]. Their aroma can be more or less intense depending on their content of molecules that provide a potential biological effect, making the hydrolates useful for the food industry as preserving and/or aromatic agents [31]. In this study, for the first time, to better describe the vapor phase chemical profiles of Hys and EOs obtained from flowers and inflorescences of *R. officinalis* and inflorescences of *L. angustifolia* growing in Tuscany, we used the automated headspace sampler directly coupled with gas chromatography–mass spectrometry (HS/GC–MS) [32,33]. This sampling technique is conservative and non-destructive and does not require the use of solvent for the extraction process, thus avoiding a possible loss of components. The chemical composition of the liquid phase of EOs was also characterized by GC/MS and their antiproliferative, antibacterial and antioxidant activity was evaluated. Moreover, this is the first report revealing the biological activity of the vapor phase and the cytotoxic activity against a neuroblastoma cell line of *Lavandula angustifolia* and *Rosmarinus officinalis* Hys.

2. Materials and Methods

2.1. Materials

EOs and Hys from flowers and inflorescences of *R. officinalis* L. and inflorescences of *L. angustifolia* Mill. growing in Tuscany (San Donato in Poggio and Roccastrada, respectively), Italy and obtained by steam distillation, were directly provided by "èssenziale" Azienda Agricola, San Donato in Poggio (FI), Italy. The dates of collection of plants were: June 2020 for *R. officinalis* and July 2020 for *L. angustifolia*. Lysogeny Broth (LB) with agar, thiazolyl blue tetrazolium bromide (MTT), vinblastine sulfate, methanol, 2,2-diphenyl-1-picrylhydrazyl (DPPH), 6-hydroxy-2,5,7,8-tetramethylchroman-2-carboxylic acid (Trolox), 2,2′-azinobis (3-ethylbenzothiazoline-6-sulfonic acid) diammonium salt (ABTS) and potassium persulfate ($K_2S_2O_8$) were from Merck (Darmstadt, Germany). Gentamicin sulfate was bought from Biochrom PAN-Bio-Tech GmbH (Aidenbach, Germany).

2.2. GC–FID and GC–MS Analysis

The GC–MS analyses were performed with a gas chromatograph equipped with a flame ionization detector (FID) and coupled to a mass spectrometer (MS), Perkin Elmer Clarus 500 model (Waltham, MA, USA). The GC capillary column was a Varian Factor Four VF-1 and helium served as a carrier gas at a flow rate of 1 mL/min. The injector temperature was 280 °C and the oven temperature program started from 60 up to 220 °C for 20 min at a rate of 6 °C min^{-1}. For liquid injections, the solutions were prepared by diluting 1 μL of each EO with 1 mL of methanol and 1 μL of the sample was injected. MS operative conditions were: ionization voltage of 70 eV and acquisition mass range 40–450. Ion source and the connection parts temperature was 220 °C. The GC–TIC mass spectra were obtained by the TurboMass data analysis software (Perkin Elmer-Vers. 6.1.0). The identification of components was performed by matching their mass spectra with the spectrometer database of the NIST and Wiley libraries and comparison of their linear retention indices (LRIs) calculated against a mixture of *n*-alkanes (C8–C30). The relative average percentages of compounds were calculated by peak area normalization from GC–FID chromatograms without the use of an internal standard or correction factors. All analyses were conducted in triplicate.

2.3. HS/GC–FID and HS/GC–MS Analysis

To describe the vapor phase profile of EOs and Hys, a Perkin-Elmer Headspace (HS) Turbomatrix 40 (Waltham, MA, USA) autosampler connected to GC–MS was used [34,35]. About 1 mL of EO and 2 mL of Hy were placed separately in 20 mL vials sealed with headspace (PolyTetrafluoroethylene-PTFE)-coated silicone rubber septa and caps. The operative optimized conditions were: the sample was heated at 60 (EOs) and 80 °C (Hys) for 20 min thermostating time and an injection volume of about 10 mL (vapor phase) was sent to the capillary column of GC by a transfer line maintained at 200 °C. Quantification and identification of compounds was carried out by GC–FID and GC–MS analyses.

2.4. Cell Culturing

To assess the biological activity of the examined EOs and related Hys, human neuroblastoma SHSY5Y (ATCC® CRL-2266™) cell lines were used. The cells were maintained in a 75 cm^2 flask containing DMEM-F12 (Dulbecco's modified Eagle's medium: nutrient mixture F-12) culture medium supplemented with 10% of FBS (fetal bovine serum), 1% glutamine and 1% penicillin/streptomycin, and maintained at 37 °C with 5% CO_2 and controlled humidity. Once the cells reached confluence, they were passed into new culture vessels in a 1:20 ratio and the medium was changed every 3 days.

2.5. Cytotoxicity Test (MTT)

The cell viability of SHSY5Y cells treated with *R. officinalis* and *L. angustifolia* EOs and Hys was evaluated by MTT assay. The mitochondrial dehydrogenase activity of the control and treated cells, which reflects their cell viability, was analyzed both in a dose-

and time-dependent manner. A total of 2×10^4 cells/well were seeded in a 96-well plate 24 h before being treated. EOs were dissolved in Dimethyl Sulfoxide (DMSO) (50% *v/v*). Twelve two-fold diluted concentrations were applied from 1×10^{-1}% to 2×10^{-4}% *v/v* for the EOs and from 50% to 1×10^{-1}% *v/v* for the Hys; the DMSO (0.05% final concentration) and double distilled water (ddH_2O) were used as solvent controls. Vinblastine sulfate (Merck KGaA, Darmstadt, Germany) was used as positive control. After 24 h, 48 h and 72 h of treatment, the medium was removed, MTT solution (0.5 mg/mL) was added to the cells and they were incubated for 3 h in dark conditions at 37 °C. DMSO was used to solubilize the formazan crystals and the absorbance was read by Tecan Sunrise™ (Tecan Group Ltd., Männedorf, Switzerland) UV-vis spectrophotometer at 595 nm. The obtained optical density values were converted into percentage of cell viability and the data were elaborated with AAT Bioquest EC_{50} Calculator (Sunnyvale, CA, USA) [19] in order to obtain the concentration at which the investigated substance exerts half of its maximal response values (EC_{50}). The values were repeated three times and reported as mean ± SD.

2.6. Antibacterial Activity

To delineate the antibacterial profiles of *R. officinalis* and *L. angustifolia* EOs and Hys, the MIC (minimal inhibitory concentration), the MBC (minimal bactericidal concentration), the agar disk-diffusion method and the VPT (vapor phase test) were used. Five different bacteria strains were considered for this study: *Escherichia coli* ATCC® 25922, *Pseudomonas fluorescens* ATCC® 13525 and *Acinetobacter bohemicus* DSM 102855 as Gram-negative bacteria, and *Kocuria marina* DSM 16420 and *Bacillus cereus* ATCC® 10876 as Gram-positive bacteria. After 24 h of culturing in lysogeny broth (LB) agar, at 26 (for *P. fluorescens*, *A. bohemicus* and *B. cereus*) and 37 °C (for *K. marina* and *E. coli*), the bacterial strains were collected and used for antimicrobial assays.

2.6.1. Minimum Inhibitory Concentration (MIC) and Minimum Bactericidal Concentration (MBC)

The microwell dilution method was carried out to take over the minimum inhibitory concentration (MIC) of the EOs and their corresponding Hys. Twelve two-fold dilutions from 6.25% to 3×10^{-3}% *v/v* and from 50% to 1×10^{-1}% *v/v* for EOs and Hys, respectively, were used. Gentamicin was used as a positive control and the negative and solvent controls were also added. A total of 10^6 colony-forming unit (CFU)/mL of bacteria in LB broth was seeded in 96-well plates, and after 24 h of incubation with the treatments the bacterial growth was evaluated by adding 10 µL of MTT (200 µg/mL) to each well [36,37]. Before adding MTT in the microwell plates, 10 µL of the last four dilutions without bacterial growth were taken and seeded on LB agar Petri plates to evaluate the minimum bactericidal concentration (MBC) or the concentrations for which no bacterial growth was observed after 24 h of incubation. Furthermore, the MBC/MIC ratios were reported and the values >4 defined bacteriostatic activity, while the ratio MBC/MIC ≤4 defined bactericidal activity for the tested samples [38]. All the assays were carried out in triplicate.

2.6.2. Agar Disk-Diffusion Method and Vapor Phase Test (VPT)

The agar disk-diffusion method was carried out to study the activity of the liquid phase. A total of 10^8 CFU/mL of bacteria was seeded on a Petri dish with LB agar and 10 µL of pure EOs and 15 µL of Hys. Gentamicin (10 mg/mL) was used as a positive control. After 24 h of incubation at the corresponding temperature, the diameter of the growth inhibition halo or inhibition zone (IZ) was measured by a vernier caliper rule [36]. To evaluate the antibacterial activity of the vapor phase of the EOs and Hys, the vapor phase test (VPT) was used. The assay was performed by seeding 10^8 CFU/mL of bacteria in a Petri dish with LB agar and pouring 5 mL of LB agar in the Petri plate cover where 6 mm sterile disks had been placed and soaked with 10 and 15 µL of the EOs and the corresponding Hys, respectively. The Petri plates were sealed with parafilm in order to prevent any vapor leakage. After 24 h of incubation, the inhibition halo was measured by a vernier caliper rule [33]. The means and standard deviations were obtained by triplicate measures of the Agar disk-diffusion and VPT halos.

2.7. Antioxidant Activity

2.7.1. DPPH Scavenging Activity Assay

Using the protocol described by Sanchez-Moreno et al. [39], the antioxidant activity of *R. officinalis* and *L. angustifolia* EOs and Hys was assayed by exploiting the 1,1-diphenyl-2-picrilidrazil radical (DPPH•) properties in the DPPH scavenging activity assay. A total of 100 μL of twelve concentrations obtained by geometric dilutions of each sample (from 25% to 0.01%) were added to a 96-well plate and mixed with 100 μL of a solid DPPH methanol solution (0.2 mM). EOs and Hys without DPPH solution were added as controls and Trolox dilutions and DPPH plus methanol were also added. After 30 min of incubation, the absorbances were measured by a Tecan Sunrise™ UV-vis spectrophotometer (517 nm). Three replicates of the experiment were carried out.

2.7.2. ABTS Radical Scavenging Assay

The ABTS (2,2′-azino-bis(3-ethylbenzothiazoline-6-sulfonic acid) diammonium salt) assay was used to investigate the radical scavenging activity [40]. An aqueous 7 mM solution of ABTS was mixed with $K_2S_2O_8$ (140 mM) and incubated for 16 h, protected from light at room temperature to allow compounds to form the ABTS+• radical cation. The obtained solution was then diluted in ethanol until the absorbance of 0.70 ± 0.02 at 734 nm was reached. Five geometrical dilutions of each sample were prepared and 20 μL was added to 980 μL of the ABTS+• solution (starting from 2% to 0.125%). After 5 min of incubation, the absorbances of the resulting solutions were measured using a Jasco (Jasco Corporation, Lecco, Italy) V-630 UV-Visible spectrophotometer at 734 nm by the Spectra Manager™ software (version II). The blank consisted of a solution composed of 20 μL of geometrical dilutions of EOs and Hys and 980 μL of ethanol. Trolox geometrical dilutions were used as a positive control. Data were collected in triplicate.

2.8. IC50 and Trolox Equivalents

Using the following equation, the percentage of antioxidant activity ($AA\%$) was estimated:

$$AA\% = \left[\frac{A_{blank} - A_{sample}}{A_{blank}}\right] \times 100 \tag{1}$$

The percentage of antioxidant activity ($AA\%$) (y) was then plotted against the sample concentrations (x) to form a regression line ($y = mx + q$). The IC_{50} value was calculated by the following formula:

$$IC_{50} = \frac{50 - q}{m} \tag{2}$$

The Trolox equivalent antioxidant capacity (TEAC) was also used to express the antioxidant capacity and expressed in μmol Trolox/mg of EOs or Hys.

2.9. Statistical Analysis

The Friedman test, a well-known distribution-free test for the randomized blocks analysis of variance [41], was used for investigating homogeneity in the cytotoxic activity of EOs and Hys across time. Furthermore, for both cytotoxic and antibacterial activities, statistical discrepancies between *R. officinalis* and *L. angustifolia* were evaluated by adopting the Mann–Whitney–Wilcoxon nonparametric test [41]. All the statistical tests were performed using R Statistical Software version 4.1.0 (The R foundation for Statistical Computing, Vienna, Austria).

3. Results

3.1. Liquid and Vapor Phase EOs Chemical Volatile Composition

GC–MS and HS/GC–MS techniques were used to investigate the liquid- and vapor-phase chemical composition of *R. officinalis* and *L. angustifolia* EOs and Hys. The composition of *R. officinalis* EO in Table 1 is reported. Monoterpene hydrocarbons (71.5% and 89.5%) were the main compounds followed by oxygenated monoterpenes (26.0% and 10.5%) in the liquid and vapor phase, respectively. Among them, α-pinene (51.2%; 74.7%) and 1,8-cineole (20.1%; 10.0%) were the most abundant components in both phases. Sesquiterpene compounds such as β-caryophyllene (1.4%), α-curcumene (0.1%) and caryophyllene oxide (0.1%) were detected only in the liquid phase.

Table 1. Chemical composition (%) of liquid and vapor phase *R. officinalis* essential oil (EO).

N°	COMPONENT [1]	LRI [2]	LRI [3]	*R. officinalis* (%) [4]	*R. officinalis* (%) [5]
1	α-pinene	945	943	51.2 ± 0.02	74.7 ± 0.1
2	camphene	948	946	5.6 ± 0.04	6.5 ± 0.02
3	dehydrosabinene	961	958	1.1 ± 0.04	1.1 ± 0.02
4	β-myrcene	982	983	2.0 ± 0.05	1.1 ± 0.02
5	β-pinene	990	986	3.4 ± 0.03	2.8 ± 0.02
6	α-phellandrene	1000	1006 *	0.2 ± 0.02	-
7	α-terpinene	1010	1008	0.4 ± 0.02	0.2 ± 0.02
8	p-cymene	1020	1016	2.0 ± 0.05	0.9 ± 0.02
9	limonene	1022	1023	4.8 ± 0.03	1.7 ± 0.02
10	1,8-cineole	1025	1027	20.1 ± 0.15	10.0 ± 0.05
11	γ-terpinene	1051	1054	0.9 ± 0.04	0.3 ± 0.02
12	terpinolene	1080	1078	0.7 ± 0.02	0.2 ± 0.02
13	p-cymenene	1085	1083.4	0.1 ± 0.01	-
14	linalool	1096	1092	1.0 ± 0.05	0.2 ± 0.02
15	camphor	1125	1126	0.8 ± 0.06	0.2 ± 0.04
16	borneol	1155	1152	0.2 ± 0.02	0.1 ± 0.01
17	endo-borneol	1158	1155	0.9 ± 0.03	-
18	terpinen-4-ol	1161	1160	0.3 ± 0.02	-
19	α-terpineol	1182	1183	0.4 ± 0.01	-
20	verbenone	1192	1196	0.6 ± 0.02	-
21	geraniol	1234	1237	0.4 ± 0.02	-
22	bornyl acetate	1262	1268	1.0 ± 0.07	-
23	nerol acetate	1362	1363	0.3 ± 0.02	-
24	β-caryophyllene	1424	1426	1.4 ± 0.03	-
25	α-curcumene	1478	1475	0.1 ± 0.02	-
26	caryophyllene oxide	1586	1583	0.1	-
	SUM (%)			100.0	100.0
	Monoterpene hydrocarbons			71.5	89.5
	Oxygenated monoterpenes			26.0	10.5
	Sesquiterpene hydrocarbons			1.5	-
	Oxygenated sesquiterpene			0.1	-
	Others			-	-

[1] The components are reported according to their elution order on a polar column; [2] linear retention indices measured on a polar column; [3] linear retention indices from literature; [4] percentage mean values of *R. officinalis* EO components (liquid phase); [5] percentage mean values ± SD (standard deviation) of *R. officinalis* EO components (vapor phase); * Normal Alkane retention index; -, not detected.

Twenty-five volatile compounds were identified in *L. angustifolia* EO and they are listed in Table 2. In this EO, oxygenated monoterpenes prevailed over monoterpene hydrocarbons with relative percentages equal to 86.9% in the liquid phase and 53.7% in the vapor phase. Linalool (49.9% and 26.2%) was the major compound both in the liquid and vapor phase, respectively. The second most abundant compound was linalyl acetate (17.9%) in the liquid phase while α-pinene (17.8%) was in the vapor phase. Additionally, in this case, sesquiterpenes were present only in the liquid phase with β-caryophyllene (1.5%) and α-farnesene (1.2%) as main components.

Table 2. Chemical composition (%) of liquid and vapor phase *L. angustifolia* EO.

N°	COMPONENT [1]	LRI [2]	LRI [3]	*L. angustifolia* (%) [4]	*L. angustifolia* (%) [5]
1	α-pinene	945	943	1.6 ± 0.02	17.8 ± 0.02
2	camphene	948	946	0.8 ± 0.02	8.9 ± 0.05
3	β-pinene	990	986	2.3 ± 0.02	1.2 ± 0.02
4	α-phellandrene	1000	1006 *	0.1 ± 0.02	0.5 ± 0.05
5	α-terpinene	1010	1008	0.2 ± 0.02	0.5 ± 0.03
6	p-cymene	1020	1016	0.1 ± 0.02	1.0 ± 0.02
7	limonene	1022	1023	1.2 ± 0.02	6.2 ± 0.02
8	1,8-cineole	1025	1027	5.7 ± 0.02	18.0 ± 0.03
9	cis-β-ocimene	1033	1032	0.1 ± 0.02	3.2 ± 0.05
10	trans-β-ocimene	1041	1043	1.4 ± 0.02	3.5 ± 0.03
11	γ-terpinene	1051	1054	0.6 ± 0.02	2.3 ± 0.03
12	linalol oxide	1073	1073	0.5 ± 0.02	0.2 ± 0.03
13	terpinolene	1080	1078	0.6 ± 0.01	1.1 ± 0.03
14	linalool	1096	1092	49.9 ± 0.14	26.2 ± 0.05
15	camphor	1125	1126	3.2 ± 0.04	6.9 ± 0.05
16	borneol	1155	1152	3.9 ± 0.02	0.7 ± 0.02
17	terpinen-4-ol	1161	1160	5.0 ± 0.05	1.4 ± 0.02
18	α-terpineol	1182	1183	0.8 ± 0.04	0.1 ± 0.02
19	linalyl acetate	1251	1252	17.9 ± 0.02	0.2 ± 0.02
20	β-caryophyllene	1424	1426	1.5 ± 0.02	-
21	cis-β-farnesene	1444	1441	0.8 ± 0.01	-
22	β-bisabolene	1500	1501	0.3 ± 0.02	-
23	α-farnesene	1505	1506	1.2 ± 0.02	-
24	caryophyllene oxide	1586	1583	tr	-
25	α-bisabolol	1662	1665	0.3 ± 0.02	-
	SUM (%)			100.0	99.9
	Monoterpene hydrocarbons			9.0	46.2
	Oxygenated monoterpenes			86.9	53.7
	Sesquiterpene hydrocarbons			3.8	-
	Oxygenated sesquiterpene			0.3	-
	Others			-	-

[1] The components are reported according to their elution order on a polar column; [2] linear retention indices measured on a polar column; [3] linear retention indices from literature; [4] percentage mean values of *L. officinalis* EO components (liquid phase); [5] percentage mean values ± SD (standard deviation) of *L. officinalis* EO components (vapor phase); * Normal Alkane retention index; -, not detected; tr, trace < 0.1.

3.2. Vapor Phase Hys Chemical Composition

The volatile chemically identified compounds of Hys vapor phase are listed in Table 3. They were only composed of oxygenated monoterpenes, and 1,8-cineole (56.2%) and linalool (42.9%) were the major components of *R. officinalis* and *L. officinalis* Hys, respectively. Camphor (20.3% and 18.4%) was the second major component with similar percentage values in both Hys.

Table 3. Chemical composition (%) of vapor-phase *R. officinalis and L. angustifolia* Hys.

N°	COMPONENT [1]	LRI [2]	LRI [3]	*R. officinalis* (%) [4]	*L. angustifolia* (%) [5]
1	1,8-cineole	1025	1027	56.2 ± 0.04	11.8 ± 0.03
2	linalol oxide	1073	1073	-	0.1 ± 0.01
3	linalool	1096	1092	4.2 ± 0.02	42.9 ± 0.05
4	camphor	1125	1126	20.3 ± 0.02	18.4 ± 0.02
5	borneol	1155	1152	10.6 ± 0.02	5.8 ± 0.02
6	terpinen-4-ol	1161	1160	1.6 ± 0.02	8.4 ± 0.02
7	α-terpineol	1182	1183	2.0 ± 0.04	12.6 ± 0.02
8	verbenone	1192	1196	5.1 ± 0.09	-
	SUM (%)			100.0	100.0

Table 3. *Cont.*

N°	COMPONENT [1]	LRI [2]	LRI [3]	*R. officinalis* (%) [4]	*L. angustifolia* (%) [5]
	Monoterpene hydrocarbons			-	-
	Oxygenated monoterpenes			100.0	100.0
	Sesquiterpene hydrocarbons			-	-
	Oxygenated sesquiterpene			-	-
	Others			-	-

[1] The components are reported according to their elution order on a polar column; [2] linear retention indices measured on a polar column; [3] linear retention indices from literature; [4] percentage mean values ± SD (standard deviation) of *R. officinalis* Hydrolate (Hy) components; [5] percentage mean values of *L. officinalis* Hy components; -, not detected.

The distribution of the main volatile compounds detected in *R. officinalis* and *L. officinalis* Hys and in the vapor-phase EOs is shown in Figures 1 and 2.

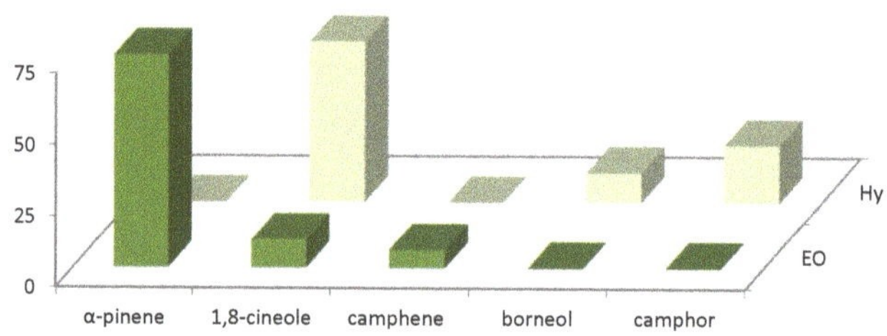

Figure 1. Bar graph of main compounds detected in vapor-phase *R. officinalis* Hy and EO.

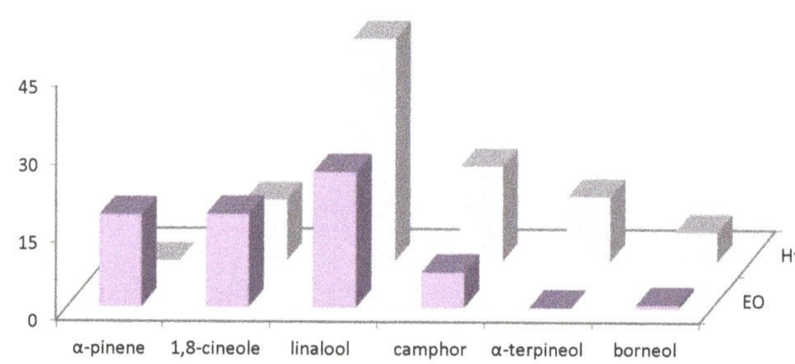

Figure 2. Bar graph of main compounds detected in vapor-phase *L. angustifolia* Hy and EO.

3.3. Cytotoxic Activity

To define the cytotoxic effects of *R. officinalis* and *L. angustifolia* EOs and their corresponding Hys, MTT assays were carried out and the EC_{50} values after 24, 48 and 72 h for SHSY5Y cells were reported (Figure 3). Friedman test *p*-values for assessing the homogeneity of EC_{50} across time with respect to EOs and Hys of both plant species are displayed in Table 4. A very mild significance for Hys (*p*-value equal to 5%) can be detected, while, as highlighted also in Figure 3a, EOs were revealed to be cytotoxic in a time-independent manner, showing EC_{50} values quite similar in the two plant species considered, from 0.05% to 0.06% v/v for the three time points evaluated (*p*-values higher than 5%). Even if *p*-values related to Hys are equal to 5%, the corresponding hypothesis could be barely rejected due to the rather small sample size. Consequently, the Hys EC_{50} values may be considered substantially stable for each time point and plant species. *R. officinalis* Hys was slightly lower at 24 h than the *L. angustifolia* Hys (26.82 ± 2.39% and 30.18 ± 1.11%, respectively) while the latter was more active (19.96 ± 4.7% and 12.78 ± 0.58%) after 48 h and after 72 h (21.53 ± 3.28% and 11.72 ± 0.60%), respectively (Figure 3b). An EC_{50} value of 1.94 ± 0.18 nM for vinblastine treatment was obtained. DMSO control did not affect cell viability.

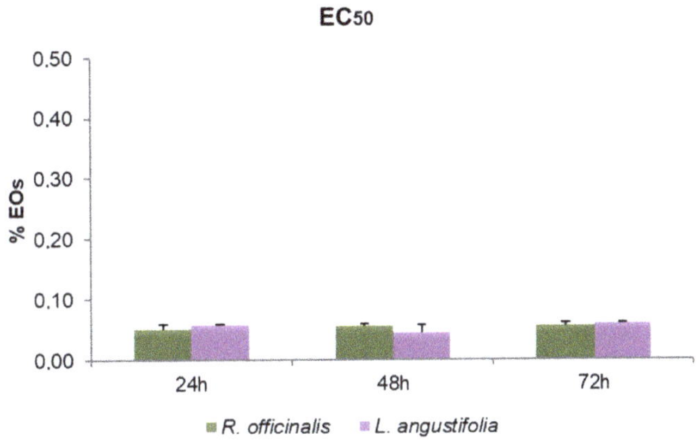

(a)

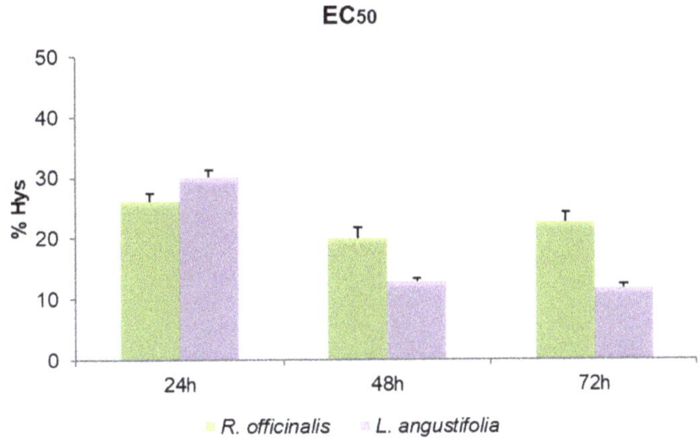

(b)

Figure 3. Bar graphs of EC_{50} values obtained by thiazolyl blue tetrazolium bromide assay (MTT) after 24, 48 and 72 h of SHSY5Y cell treatments, (**a**) with *R. officinalis* and *L. angustifolia* EOs and (**b**) with the corresponding Hys. Error bars: standard deviation.

Table 4. Friedman test *p*-values for testing homogeneity across time.

Sample	*p*-Value
R. officinalis EOs	0.717
L. angustifolia EOs	0.097
R. officinalis Hys	0.050
L. angustifolia Hys	0.050

Owing to the effective homogeneity resulting from the Friedman test, the data were reduced to the means of EC_{50} measurements across time. Therefore, in order to assess the EC_{50} values homogeneity between *R. officinalis* and *L. angustifolia*, two Mann–Whitney–Wilcoxon tests were implemented. From the obtained results (both *p*-values equal to 1), *R. officinalis* and *L. angustifolia* produced statistically equivalent EC_{50} values both for EOs and Hys.

3.4. Antibacterial Activity

Investigations on *R. officinalis* and *L. angustifolia* EOs and Hys antibacterial activity were executed by different assays. In Table 5, the results obtained by *R. officinalis* EO are listed. The highest antibacterial activity was against *A. bohemicus* with an MIC value of 0.19%, MBC value of 0.39%, IZ of 9.17 ± 0.76 mm and VIZ of 80.00 ± 00 mm. Concerning the other tested bacterial strains, MIC values ranged from 0.39% to 3.13% and MBC values from 0.39% to 6.25%. The MIC/MBC ratio showed that the tested EOs possessed bactericidal properties. IZs were 7.33 ± 0.58 and 8.33 ± 1.52 mm for *K. marina* and *B. cereus*, respectively, and VIZs were 80.00 ± 00 mm. For *E. coli*, IZ was 7.00 ± 0.00 mm, while the EO vapor phase did not determine bacterial growth inhibition. *R. officinalis* Hy was not active against growth of the tested bacteria (Table 5).

Table 5. Antibacterial activity of *Rosmarinus officinalis* EO and Hy.

	R. officinalis EO					*R. officinalis* Hy				
	MIC [1]	MBC [2]	MBC/MIC Ratio	IZ [3]	VIZ [4]	MIC [1]	MBC [2]	MBC/MIC Ratio	IZ [3]	VIZ [4]
E. coli	3.13	3.13	1.00	7.00 ± 0.00	-	na	na	-	-	-
P. fluorescens	3.13	6.25	0.50	-	-	na	na	-	-	-
A. bohemicus	0.19	0.39	0.50	9.17 ± 0.76	80 ± 00	na	na	-	-	-
K. marina	1.56	3.13	0.50	7.33 ± 0.58	80 ± 00	na	na	-	-	-
B. cereus	0.39	0.39	1.00	8.33 ± 1.52	80 ± 00	na	na	-	-	-

[1] Minimal inhibitory concentration (% *v*/*v*) of *R. officinalis* EO and HY; [2] minimal bactericidal concentration (% *v*/*v*) of EO and Hy; [3] growth inhibition zone by disc diffusion assay (mm); [4] growth inhibition zone by vapor phase test (mm); na, not attained; -, not detected. Values are expressed as means ± SD. Gentamicin determined MIC and MBC values of 6.25 µg/mL for *E. coli*, 3.13 µg/mL for *P. fluorescens*, 6.25 µg/mL for *A. bohemicus*, 1.56 µg/mL for *K. marina* and 1.56 µg/mL for *B. cereus* and IZ values of 17.00 ± 1 mm for *E. coli*, 19.67 ± 0.58 mm for *P.'fluorescens*, 24.33 ± 1.53 mm for *A. bohemicus*, 24.67 ± 1.53 mm for *K. marina* and 19.67 ± 1.53 mm for *B. cereus*.

L. angustifolia EO presented a high antibacterial activity with MIC values of 0.39%, 1.56%, 0.19%, 0.78% and 0.19% and MBC values of 0.39%, 3.13%, 0.39%, 0.78% and 0.19% for *E. coli*, *P. fluorescens*, *A. bohemicus*, *K. marina* and *B. cereus*, respectively. MIC/MBC ratios showed that the tested EO possessed bactericidal properties. IZ values for *R. officinalis* Hy were 11.00 ± 1.00, 7.17 ± 0.76, 11.67 ± 1.15, 11.33 ± 1.53 and 10.67 ± 0.58 mm against *E. coli*, *P. fluorescens*, *A. bohemicus*, *K. marina* and *B. cereus*. VIZ values were 6.17 ± 1.04 and 0.67 ± 1.15 mm against *A. bohemicus* and *B. cereus*, respectively. *L. angustifolia* Hy was not effective against the tested bacterial strains growth (Table 6).

Table 6. Antibacterial activity of *L. angustifolia* EO and Hy.

	L. angustifolia EO					*L. angustifolia* Hy				
	MIC [1]	MBC [2]	MBC/MIC Ratio	IZ [3]	VIZ [4]	MIC [1]	MBC [2]	MBC/MIC Ratio	IZ [3]	VIZ [4]
E. coli	0.39	0.39	1.00	11.00 ± 1.00	-	na	na	-	-	-
P. fluorescens	1.56	3.13	0.50	7.17 ± 0.76	-	na	na	-	-	-
A. bohemicus	0.19	0.39	0.50	11.67 ± 1.15	6.17 ± 1.04	na	na	-	-	-
K. marina	0.78	0.78	1.00	11.33 ± 1.53	-	na	na	-	-	-
B. cereus	0.19	0.19	1.00	10.67 ± 0.58	0.67 ± 1.15	na	na	-	-	-

[1] Minimal inhibitory concentration (% *v*/*v*) of *L. angustifolia* EO and HY; [2] minimal bactericidal concentration (% *v*/*v*) of EO and Hy; [3] growth inhibition zone by disc diffusion assay (mm); [4] growth inhibition zone by vapor phase test (mm); na, not attained; -, not detected. Values are expressed as means ± SD. Gentamicin determined MIC and MBC values of 6.25 µg/mL for *E. coli*, 3.13 µg/mL for *P. fluorescens*, 6.25 µg/mL for *A. bohemicus*, 1.56 µg/mL for *K. marina* and 1.56 µg/mL for *B. cereus* and IZ values of 17.00 ± 1 mm for *E. coli*, 19.67 ± 0.58 mm for *P. fluorescens*, 24.33 ± 1.53 mm for *A. bohemicus*, 24.67 ± 1.53 mm for *K. marina* and 19.67 ± 1.53 mm for *B. cereus*.

IZ and VIZ values for *R. officinalis* and *L. angustifolia* EOs were considered for assessing a significant discrepancy in the antibacterial activities. Mann–Whitney–Wilcoxon p-values suggest that, for any significance level greater than 5%, the antibacterial activity against

E. coli, *A. bohemicus* and *K. marina*, in terms of IZ values, is greater for *L. angustifolia* EO than for *R. officinalis* EO (exact p-values equal to 5%), while this cannot be concluded for the antibacterial activity against *B. cereus* (exact *p*-value equal to 10%). Contrarily, when VIZ values are considered, the Mann–Whitney–Wilcoxon p-values suggest that for any significance level greater than 5%, the antibacterial activity against *A. bohemicus* and *B. cereus* is greater for *R. officinalis* EO than for *L. angustifolia* EO (exact *p*-values equal to 5%).

3.5. Antioxidant Activity of *R. officinalis* and *L. angustifolia* EOs and Hys

The antioxidant potential of the EOs and the Hys was investigated by DPPH and ABTS assays reporting the IC_{50} values (Table 7). Results are also expressed in μM Trolox/mg of samples. The highest antioxidant activity was measured in the EOs followed by Hys of both plants. For *R. officinalis* EO, IC_{50} was 13.48 ± 1.58 and 20.21 ± 2.72 μg/mL, while for *L. angustifolia* EO IC_{50} was 7.75 ± 0.10 and 18.71 ± 2.15 μg/mL for DPPH and ABTS, respectively. TEAC values were 1.90 ± 1.14 and 23.53 ± 2.43 μmol/mg for *R. officinalis* EO and 3.30 ± 0.09 and 25.45 ± 3.72 μmol/mg for *L. angustifolia* EO, for DPPH and ABTS, respectively. *R. officinalis* and *L. angustifolia* Hys showed antioxidant activity lower than the corresponding EOs. In particular, *R. officinalis* Hy IC_{50} was 136.30 ± 3.85 and 349.42 ± 19.32 μg/mL and TEAC values were 0.22 ± 0.02 and 1.35 ± 0.03 μmol/mg by DPPH and ABTS, respectively. On the contrary, *L. angustifolia* Hy antioxidant activity was slightly lower than *R. officinalis* Hy in DPPH assay (240.02 ± 13.65 μg/mL and 0.12 ± 0.01 μmol/mg) and higher in ABTS assay (181.24 ± 15.71 μg/mL and 2.62 ± 0.30 μmol/mg, for IC_{50} and TEAC, respectively).

Table 7. Antioxidant activity of *R. officinalis* and *L. angustifolia* EOs and Hys.

		***R. officinalis* EO**	***R. officinalis* Hy**	***L. angustifolia* EO**	***L. angustifolia* Hy**
DPPH	**IC_{50} ***	13.48 ± 1.59	136.30 ± 3.85	7.75 ± 0.10	240.02 ± 13.65
	TEAC **	1.90 ± 0.15	0.22 ± 0.03	3.30 ± 0.10	0.12 ± 0.01
ABTS•+	**IC_{50} ***	20.20 ± 2.72	349.42 ± 19.32	18.71 ± 2.16	181.24 ±15.71
	TEAC **	23.53 ± 2.43	1.35 ± 0.03	25.45 ± 3.73	2.62 ± 0.31

* μg/mL; ** μmol/mg. Values are expressed as means ± SD; half maximal inhibitory concentration (IC_{50}); 2,2′-azino-bis (3-ethylbenzothiazoline-6-sulfonic acid) diammonium salt based assay (ABTS•+); 2,2-diphenyl-1-picrylhydrazyl assay (DPPH); Trolox equivalent antioxidant capacity (TEAC).

4. Discussion

In our investigation, liquid and vapor phases of *L. angustifolia* EO were characterized by linalool (49.9% and 26.2%) as a major constituent, while *R. officinalis* EO was characterized by α-pinene (51.2% and 74.7%). The chemical profile of *L. angustifolia* and *R. officinalis* EOs obtained by aerial parts of plants growing in Syria was investigated by Al-Younis et al. [42], showing a similar composition with borneol (16.25%) and linalool (35.12%) as the main components, respectively, whereas *L. angustifolia* EO collected in Xinjiang [43] exhibited linalyl acetate (28.89%) as a principal molecule, as well as the Himalayan one (47.56%) [44]. Reagrding *R. officinalis* EO, 1,8-cineole was the most abundant component in EO from China [45] and Belgrade [46] (26.54% and 43.77%, respectively). On the other side, α-pinene (43.9%) and p-cymene (44.02%) were found to be those with the higher percentage in *R. officinalis* EO from Iran [47] and Turkey [48]. The present study is the first on the characterization of the aroma profiles of *R. officinalis* and *L. angustifolia* Hys grown in the Tuscany region using HS/GC–MS. The obtained results highlight that they were characterized exclusively by oxygenated monoterpenes, among which 1,8-cineole (56.2% in *R. officinalis* Hy) and linalool (42.9% in *L. angustifolia* Hy) were the major exponents. Previous studies showed a different chemical composition, with camphor (24.9%), terpinen-4-ol (51.9%) and verbenone (45.31%) as the main components of the *R. officinalis* hydrolates from Japan [49], Colombia [50] and Korea [51], respectively. Regarding *L. angustifolia* Hy, a similar composition to that described by us was reported for extracts from Poland [27,52], in which linalool (24.6% and 26.5%, respectively) was also the major detected compound.

Of interest, *L. angustifolia* Hy from Croatia [53] was characterized by linalool (23.2%) when steam distillation was used to obtain the extract, and by 1,8-cineole (20.6%) through hydrodistillation. In fact, the chemical profile of Hys can also vary according to the distillation method used [54].

Our findings and the cited references confirm that the variability of the chemical composition of EOs and Hys depends on various exogenous and endogenous factors, such as the area of provenance of the plant and the extraction method [55].

The biological properties of the *R. officinalis* and *L. angustifolia* EOs were investigated with in vivo and in vitro models [9,56] and numerous studies demonstrated their exertion of biological activities [57]. In the present study, *R. officinalis* and *L. angustifolia* EOs and Hys, cytotoxic, antimicrobial and antioxidant activities were investigated. Both EOs exerted cytotoxic activities with low EC_{50} values, corresponding with a high inhibition of SHSY5Y cell proliferation. Regarding Hys treatments, higher EC_{50} values were obtained and *L. angustifolia* Hy IC_{50} was slightly lower than *R. officinalis* Hy at 48 and 72 h. Studies on Hys cytotoxicity and antiproliferative effects are limited. *Melissa officinalis, Achillea teretifolia, Achillea aleppica, Origanum onites* and *Salvia fruticosa* have been investigated on a colorectal cancer cell line to determine their cytotoxic and cytostatic effects, and *O. onites* Hy was the most effective [58].

In this paper, DPPH and ABTS assays demonstrated that the EOs and Hys possess antioxidant activities, although showing different IC_{50} values. Antibacterial properties of the Hys were investigated and, unlike EOs, they were not able to inhibit bacterial growth in the different assays.

As reviewed by D'Amato et al. [31], the antibacterial and antifungal, as well antioxidant, properties of the Hys were demonstrated for different plant species, and their uses and applications could be evaluated to control microorganism growth and oxidative processes in food shelf-life. Furthermore, Hys were demonstrated to be active against biofilm production [59,60], and thus could be used as a natural antimicrobial agent for food production [54]. As reported by Šilha et al. [53], no antibacterial effect of *L. angustifolia* Hy was detected against eight strains of *Arcobacter*-like bacteria and against *Staphylococcus aureus*, *Enterococcus faecalis*, *Pseudomonas aeruginosa*, *Escherichia coli* and the yeast *Candida albicans*. On the contrary, concentrated Hys, obtained by solid-phase extraction and tested against the same microorganisms, exercised a considerable antimicrobial activity.

On the other side, *R. officinalis* EO showed antioxidant and antimicrobial properties while its corresponding Hy had neither a potential effect against *E. coli*, *P. aeruginosa*, *S. aureus* and *C. albicans* and *A. niger*, nor antioxidant properties [50].

Notably, lavandin (*L. x intermedia*) Hy was not reported to be active against *E. coli* and *B. cereus*, whereas it showed antibacterial activity against the same bacterial strains when formulated in nanoemulsion [61].

Hys being composed of the condensed water of the distillation process and by only some volatile oil components, their chemical composition is different with respect to the corresponding EOs; however, the amounts of the main components can vary greatly [31,62]. Generally, Hys exert their biological activities at high concentrations, reflecting their low terpene amount. Since they are aqueous solutions, the hydrophilic environment likely facilitates the terpenes' availability, enhancing their biological activity [54].

5. Conclusions

Chemical investigations performed by HS/GC–MS revealed that EOs and Hys of *R. officinalis* and *L. angustifolia* are rich in bioactive compounds. Both EOs revealed a good antibacterial and antioxidant activity, while their respective Hys exerted a slight antioxidant activity and were completely inactive on the selected bacterial strains. The antiproliferative activity was also evaluated by highlighting, for the first time, that not only EOs but also Hys exerted a cytotoxic effect.

In conclusion, *R. officinalis* and *L. angustifolia* EOs, thanks to their exhibited biological activities, could have potential applications in various fields, including foods and beverages.

It is also interesting to note that the results obtained with the Hys on the SHSY5Y cell line underline the potentiality of these by-products of the distillation process. In this regard, our findings will be useful for further studies and applications.

Author Contributions: Conceptualization, S.G., V.L.M. and E.O.; investigation, S.G., V.L.M. and E.O.; data curation, S.G., V.L.M., S.F. and E.O.; writing—original draft preparation, S.G., V.L.M. and E.O.; writing—review and editing, S.G. and E.O.; funding acquisition, A.T. and P.G. All the authors critically edited the manuscript before submission. All authors have read and agreed to the published version of the manuscript.

Funding: This research received no external funding.

Institutional Review Board Statement: Not applicable.

Informed Consent Statement: Not applicable.

Data Availability Statement: All generated data are included in this article.

Acknowledgments: The authors are thankful to "èssenziale" Azienda Agricola, Empoli (FI), Italy, for providing EOs and Hys from flowers and inflorescences of *L. angustifolia* and *R. officinalis*.

Conflicts of Interest: The authors declare no conflict of interest.

References

1. Jiang, T.A. Health benefits of culinary herbs and spices. *J. AOAC Int.* **2019**, *102*, 395–411. [CrossRef] [PubMed]
2. Motti, R. Wild Plants Used as Herbs and Spices in Italy: An Ethnobotanical Review. *Plants* **2021**, *10*, 563. [CrossRef] [PubMed]
3. Cocan, I.; Alexa, E.; Danciu, C.; Radulov, I.; Galuscan, A.; Obistioiu, D.; Morvay, A.A.; Sumalan, R.M.; Poiana, M.; Pop, G.; et al. Phytochemical screening and biological activity of Lamiaceae family plant extracts. *Exp. Ther. Med.* **2018**, *15*, 1863–1870. [CrossRef] [PubMed]
4. Mamadalieva, N.Z.; Akramov, D.K.; Ovidi, E.; Tiezzi, A.; Nahar, L.; Azimova, S.S.; Sarker, S.D. Aromatic medicinal plants of the Lamiaceae family from Uzbekistan: Ethnopharmacology, essential oils composition, and biological activities. *Medicines* **2017**, *4*, 8. [CrossRef]
5. de Macedo, L.M.; Santos, É.M.D.; Militão, L.; Tundisi, L.L.; Ataide, J.A.; Souto, E.B.; Mazzola, P.G. Rosemary (*Rosmarinus officinalis* L., syn *Salvia rosmarinus* Spenn.) and Its Topical Applications: A Review. *Plants* **2020**, *9*, 651. [CrossRef]
6. Ydyrys, A.; Zhaparkulova, N.; Aralbaeva, A.; Mamataeva, A.; Seilkhan, A.; Syraiyl, S.; Murzakhmetova, M. Systematic analysis of combined antioxidant and membrane-stabilizing properties of several Lamiaceae family Kazakhstani plants for potential production of tea beverages. *Plants* **2021**, *10*, 666. [CrossRef]
7. González-Minero, F.J.; Bravo-Díaz, L.; Ayala-Gómez, A. *Rosmarinus officinalis* L. (Rosemary): An Ancient Plant with Uses in Personal Healthcare and Cosmetics. *Cosmetics* **2020**, *7*, 77. [CrossRef]
8. Ribeiro-Santos, R.; Carvalho-Costa, D.; Cavaleiro, C.; Costa, H.S.; Albuquerque, T.G.; Castilho, M.C.; Ramos, F.; Melo, N.R.; Sanches-Silva, A. A novel insight on an ancient aromatic plant: The rosemary (*Rosmarinus officinalis* L.). *Trends Food Sci. Tech.* **2015**, *45*, 355–368. [CrossRef]
9. De Oliveira, J.R.; Camargo, S.E.A.; De Oliveira, L.D. *Rosmarinus officinalis* L. (rosemary) as therapeutic and prophylactic agent. *J. Biomed. Sci.* **2019**, *26*, 1–22. [CrossRef]
10. Moja, S.; Guitton, Y.; Nicolè, F.; Legendre, L.; Pasquier, B.; Upson, T.; Jullien, F. Genome size and plastid trnK-matK markers give new insights into the evolutionary history of the genus *Lavandula* L. *Plant Biosyst. Int. J. Deal. All Asp. Plant Biol.* **2015**, *150*, 1216–1224. [CrossRef]
11. Hassiotis, C.N.; Ntana, F.; Lazari, D.M.; Poulios, S.; Vlachonasios, K.E. Environmental and developmental factors affect essential oil production and quality of *Lavandula angustifolia* during flowering period. *Ind. Crop. Prod.* **2014**, *62*, 359–366. [CrossRef]
12. de Sousa, D.P.; de Almeida Soares Hocayen, P.; Andrade, L.N.; Andreatini, R. A systematic review of the anxiolytic-like effects of essential oils in animal models. *Molecules* **2015**, *20*, 18620–18660. [CrossRef] [PubMed]
13. Ayaz, M.; Sadiq, A.; Junaid, M.; Ullah, F.; Subhan, F.; Ahmed, J. Neuroprotective and anti-aging potentials of essential oils from aromatic and medicinal plants. *Front. Aging Neurosci.* **2017**, *9*, 168. [CrossRef]
14. López, V.; Nielsen, B.; Solas, M.; Ramírez, M.J.; Jäger, A.K. Exploring pharmacological mechanisms of lavender (*Lavandula angustifolia*) essential oil on central nervous system targets. *Front. Pharmacol.* **2017**, *8*, 280. [CrossRef]
15. Nikolić, M.; Jovanović, K.K.; Marković, T.; Marković, D.; Gligorijević, N.; Radulović, S.; Soković, M. Chemical composition, antimicrobial, and cytotoxic properties of five Lamiaceae essential oils. *Ind. Crop. Prod.* **2014**, *61*, 225–232. [CrossRef]
16. Gezici, S. Promising anticancer activity of lavender (*Lavandula angustifolia* Mill.) essential oil through induction of both apoptosis and necrosis. *Ann. Phytomed.* **2018**, *7*, 38–45. [CrossRef]
17. Garzoli, S.; Turchetti, G.; Giacomello, P.; Tiezzi, A.; Laghezza Masci, V.; Ovidi, E. Liquid and vapour phase of lavandin (*Lavandula* × *intermedia*) essential oil: Chemical composition and antimicrobial activity. *Molecules* **2019**, *24*, 2701. [CrossRef] [PubMed]

18. Boukhatem, M.N.; Sudha, T.; Darwish, N.H.E.; Chader, H.; Belkadi, A.; Rajabi, M.; Houche, A.; Benkebailli, F.; Oudjida, F.; Mousa, S.A. A New eucalyptol-rich lavender (*Lavandula stoechas* L.) essential oil: Emerging potential for therapy against inflammation and cancer. *Molecules* **2020**, *25*, 3671. [CrossRef]
19. Laghezza Masci, V.; Ovidi, E.; Taddei, A.R.; Turchetti, G.; Tiezzi, A.; Giacomello, P.; Garzoli, S. Apoptotic effects on HL60 human leukaemia cells induced by lavandin essential oil treatment. *Molecules* **2020**, *25*, 538. [CrossRef] [PubMed]
20. Abd Rashed, A.; Rathi, D.-N.G.; Ahmad Nasir, N.A.H.; Abd Rahman, A.Z. Antifungal properties of essential oils and their compounds for application in skin fungal infections: Conventional and nonconventional approaches. *Molecules* **2021**, *26*, 1093. [CrossRef]
21. Aprotosoaie, A.C.; Gille, E.; Trifan, A.; Luca, V.S.; Miron, A. Essential oils of Lavandula genus: A systematic review of their chemistry. *Phytochem. Rev.* **2017**, *16*, 761–799. [CrossRef]
22. Lawrence, B.M. Progress in essential oils, lavender oils. *Perf. Flav.* **1993**, *18*, 58–61.
23. Cseke, L.J.; Kirakosyan, A.; Kaufman, P.B.; Warber, S.L.; Duke, J.A.; Brielmann, H.L. *Natural Products from Plants*, 2nd ed.; CRC Press: Boca Raton, FL, USA, 2006.
24. Bakkali, F.; Averbeck, S.; Averbeck, D.; Idaomar, M. Biological effects of essential oils—A review. *Food Chem. Toxicol.* **2008**, *46*, 446–475. [CrossRef]
25. Dhifi, W.; Bellili, S.; Jazi, S.; Bahloul, N.; Mnif, W. Essential oils' chemical characterization and investigation of some biological activities: A critical review. *Medicines* **2016**, *3*, 25. [CrossRef] [PubMed]
26. Luo, W.; Du, Z.; Zheng, Y.; Liang, X.; Huang, G.; Zhang, Q.; Liu, Z.; Zhang, K.; Zheng, X.; Lin, L.; et al. Phytochemical composition and bioactivities of essential oils from six Lamiaceae species. *Ind. Crop. Prod.* **2019**, *133*, 357–364. [CrossRef]
27. Prusinowska, R.; Śmigielski, K.B. Composition, biological properties and therapeutic effects of lavender (*Lavandula angustifolia* L.). A review. *Herba Pol. J.* **2014**, *2*, 56–66. [CrossRef]
28. Begum, A.; Sandhya, S.; Shaffath, A.S.; Vinod, K.R.; Reddy, S.; Banji, D. An in-depth review on the medicinal flora *Rosmarinus officinalis* (Lamiaceae). *Acta Sci. Pol. Technol. Aliment.* **2013**, *12*, 61–73. [PubMed]
29. Edris, A.E. Identification and absolute quantification of the major water-soluble aroma components isolated from the hydrosols of some aromatic plants. *J. Essent. Oil Bear. Plants* **2009**, *12*, 155–161. [CrossRef]
30. Hamedi, A.; Moheimani, S.M.; Sakhteman, A.; Etemadfard, H.; Moein, M. An overview on indications and chemical composition of aromatic waters (hydrosols) as functional beverages in Persian nutrition culture and folk medicine for hyperlipidemia and cardiovascular conditions. *J. Evid. Based Complement. Altern.* **2017**, *22*, 544–561. [CrossRef]
31. D'Amato, S.; Serio, A.; López, C.C.; Paparella, A. Hydrosols: Biological activity and potential as antimicrobials for food applications. *Food Control.* **2018**, *86*, 126–137. [CrossRef]
32. Ovidi, E.; Laghezza Masci, V.; Zambelli, M.; Tiezzi, A.; Vitalini, S.; Garzoli, S. *Laurus nobilis*, *Salvia sclarea* and *Salvia officinalis* Essential Oils and Hydrolates: Evaluation of Liquid and Vapor Phase Chemical Composition and Biological Activities. *Plants* **2021**, *10*, 707. [CrossRef]
33. Garzoli, S.; Masci, V.L.; Caradonna, V.; Tiezzi, A.; Giacomello, P.; Ovidi, E. Liquid and Vapor Phase of Four Conifer-Derived Essential Oils: Comparison of Chemical Compositions and Antimicrobial and Antioxidant Properties. *Pharmaceuticals* **2021**, *14*, 134. [CrossRef] [PubMed]
34. Zhang, W.; Liang, X. Headspace Gas Chromatography-Mass Spectrometry for Volatile Components Analysis in Ipomoea Cairica (L.) Sweet Leaves: Natural Deep Eutectic Solvents as Green Extraction and Dilution Matrix. *Foods* **2019**, *8*, 205. [CrossRef] [PubMed]
35. Gören, A.C.; Topçu, A.; Bilsel, G.; Bilsel, M.; Wilkinson, J.M.; Cavanagh, H.M. Analysis of essential oil of Satureja thymbra by hydrodistillation, thermal desorber, and headspace GC/MS techniques and its antimicrobial activity. *Nat. Prod. Res.* **2004**, *18*, 189–195. [CrossRef] [PubMed]
36. Balouiri, M.; Moulay Sadiki, M.; Koraichi Ibnsouda, S. Methods for in vitro evaluating antimicrobial activity: A review. *J. Pharm. Anal.* **2016**, *6*, 71–79. [CrossRef] [PubMed]
37. Clinical and Laboratory Standards Institute. *M23–A3: Development of In Vitro Susceptibility Testing Criteria and Quality Control Parameters; Approved Guideline*, 3rd ed.; CLSI Document; Clinical and Laboratory Standards Institute: Wayne, PA, USA, 2008; pp. 1–68.
38. Gatsing, D.; Tchakoute, V.; Ngamga, D.; Kuiate, J.; Tamokou, J.; Nji-Nkah, B.; Tchouanguep, F.; Fodouop, S. In vitro antibacterial activity of *Crinum purpurascens* Herb. leaf extract against the Salmonella species causing typhoid fever and its toxicological evaluation. *Iran. J. Med. Sci.* **2009**, *34*, 126–136.
39. Sanchez-Moreno, C.; Larrauri, J.A.; Saura-Calixto, F. A procedure to measure the antiradical efficiency of polyphenols. *J. Sci. Food Agric.* **1998**, *76*, 270–276. [CrossRef]
40. Re, R.; Pellegrini, N.; Proteggente, A.; Yang, M.; Rice-Evans, C. Antioxidant activity applying an improved ABTS radical cation decolorization assay. *Free Radic. Biol. Med.* **1999**, *26*, 1231–1237. [CrossRef]
41. Hollander, M.; Douglas, A.W. *Nonparametric Statistical Methods*; John Wiley & Sons: New York, NY, USA, 1999; pp. 106–272.
42. Al-Younis, F.; Al-Naser, Z.; Al-Hakim, W. Chemical composition of *Lavandula angustifolia* Miller and *Rosmarinus officinalis* L. essential oils and fumigant toxicity against larvae of *Ephestia kuehniella* Zeller. *Int. J. Chem. Technol. Res.* **2015**, *8*, 1382–1390.
43. Xiaotian, C.; Lanyue, Z.; Chenyu, Q.; Zhiyun, D.; Peng, X.; Zhangmin, X. Chemical compositions of essential oil extracted from *Lavandula angustifolia* and its prevention of TPA-induced inflammation. *Microchem. J.* **2020**, *153*, 104458.

44. Verma, R.S.; Rahman, L.U.; Chanotiya, C.S.; Verma, R.K.; Chauhan, A.; Yadav, A.; Singh, A.; Yadav, A.K. Essential oil composition of *Lavandula angustifolia* Mill. cultivated in the mid hills of Uttarakhand, India. *J. Serb. Chem. Soc.* **2010**, *75*, 343–348. [CrossRef]
45. Jianga, Y.; Wua, N.; Fua, Y.; Wanga, W.; Luoa, M.; Zhao, C.; Zua, Y.; Liu, X. Chemical composition and antimicrobial activity of the essential oil of *Rosemary*. *Environ. Toxicol. Pharmacol.* **2011**, *32*, 63–68. [CrossRef]
46. Rašković, A.; Milanović, I.; Pavlović, N.; Ćebović, T.; Vukmirović, S.; Mikov, M. Antioxidant activity of rosemary (*Rosmarinus officinalis* L.) essential oil and its hepatoprotective potential. *BMC Complement. Altern. Med.* **2014**, *14*, 225. [CrossRef]
47. Jamshidi, R.; Afzali, Z.; Afzali, D. Chemical Composition of Hydrodistillation Essential Oil of *Rosemary* in Different Origins in Iran and Comparison with Other Countries. *Am. Eurasian J. Agric. Environ. Sci.* **2009**, *5*, 78–81.
48. Özcan, M.M.; Chalchat, J.-C. Chemical composition and antifungal activity of rosemary (*Rosmarinus officinalis* L.) oil from Turkey. *Int. J. Food Sci. Nutr.* **2008**, *59*, 691–698. [CrossRef]
49. Tomi, K.; Kitao, M.; Konishi, N.; Murakami, H.; Matsumura, Y.; Hayashi, T. Enantioselective GC–MS analysis of volatile components from rosemary (*Rosmarinus officinalis* L.) essential oils and hydrosols. *Biosci. Biotechnol. Biochem.* **2016**, *80*, 840–847. [CrossRef] [PubMed]
50. Hay, Y.-O.; Abril-Sierra, M.A.; Sequeda-Castañeda, L.G.; Bonnafous, C.; Raynaud, C. Evaluation of combinations of essential oils and essential oils with hydrosols on antimicrobial and antioxidant activities. *J. Pharm. Pharmacogn. Res.* **2018**, *6*, 216–230.
51. Hyeon Jeon, D.; Yong Moon, J.; Bong Hyun, H.; Kim Cho, S. Composition Analysis and Antioxidant Activities of the Essential Oil and the Hydrosol Extracted from *Rosmarinus officinalis* L. and *Lavandula angustifolia* Mill. Produced in Jeju. *J. Appl. Biol. Chem.* **2013**, *56*, 141–146. [CrossRef]
52. Śmigielski, K.B.; Prusinowska, R.; Krosowiak, K.; Sikora, M. Comparison of qualitative and quantitative chemical composition of hydrolate and essential oils of lavender (*Lavandula angustifolia*). *J. Essent. Oil Res.* **2013**, *25*, 291–299. [CrossRef]
53. Šilha, D.; Švarcová, K.; Bajer, T.; Královec, K.; Tesařová, E.; Moučková, K.; Pejchalová, M.; Bajerová, P. Chemical Composition of Natural Hydrolates and Their Antimicrobial Activity on Arcobacter-Like Cells in Comparison with Other Microorganisms. *Molecules* **2020**, *25*, 5654. [CrossRef] [PubMed]
54. Aćimović, M.G.; Tešević, V.V.; Smiljanić, K.T.; Cvetković, M.T.; Stanković, J.M.; Kiprovski, B.M.; Sikora, V.S. Hydrolates—by-products of essential oil distillation: Chemical composition, biological activity and potential uses. *Adv. Mater. Technol.* **2020**, *9*, 54–70.
55. Turek, C.; Stintzing, F.C. Stability of essential oils: A review. *Compr. Rev. Food Sci. Food Saf.* **2013**, *12*, 40–56. [CrossRef]
56. Benny, A.; Thomas, J. Essential oils as treatment strategy for Alzheimer's disease: Current and future perspectives. *Planta Med.* **2019**, *85*, 239–248. [PubMed]
57. Elshafie, H.S.; Camele, I. An Overview of the biological effects of some mediterranean essential oils on human health. *Biomed. Res. Int.* **2017**, *2017*, 9268468. [CrossRef] [PubMed]
58. Şimşek Sezer, E.; Bozkurt, M.; Tulukcu, E.; Uysal, T. Comparative Cytotoxic Effects of the Hydrosols of Some Ethnobotanic Plants. *Curr. Perspect. Med. Aromat. Plants (CUPMAP)* **2020**, *3*, 97–103. [CrossRef]
59. Karampoula, F.; Giaouris, E.; Deschamps, J.; Doulgeraki, A.; Nychas, G.J.E.; Dubois-Brissonnet, F. Hydrosol of *Thymbra capitata* is a highly efficient biocide against Salmonella enterica serovar typhimurium biofilm. *Appl. Environ. Microbiol.* **2016**, *82*, 5309–5319. [CrossRef] [PubMed]
60. Chorianopoulos, N.G.; Giaouris, E.D.; Skandamis, P.N.; Haroutounian, S.A.; Nychas, G.J.E. Disinfectant test agains monoculture and mixed-culture biofilms composed of technological, spoilage and pathogenic bacteria: Bactericidal effect of essential oil and hydrosol of *Satureja thymbra* and comparison with standard acid-base sanitizers. *J. Appl. Microbiol.* **2008**, *104*, 1586–1596. [CrossRef]
61. Garzoli, S.; Petralito, S.; Ovidi, E.; Turchetti, G.; Masci, V.L.; Tiezzi, A.; Trilli, J.; Cesa, S.; Casadei, M.A.; Giacomello, P.; et al. *Lavandula x intermedia* essential oil and hydrolate: Evaluation of chemical composition and antibacterial activity before and after formulation in nanoemulsion. *Ind. Crop. Prod.* **2020**, *145*, 112068. [CrossRef]
62. Inouye, S. A comparative study on the composition of forty-four hydrosols and their essential oils. *Int. J. Essent. Oil Ther.* **2008**, *2*, 89–104.

foods

MDPI

Article

An Absorption and Plasma Kinetics Study of Monoterpenes Present in Mastiha Oil in Humans

Efstathia Papada, Aristea Gioxari, Charalampia Amerikanou, Nikolaos Galanis and Andriana C. Kaliora *

Department of Dietetics and Nutritional Science, School of Health Science and Education, Harokopio University, 70 El. Venizelou Ave, 17671 Athens, Greece; efpapada@gmail.com (E.P.); arisgiox@gmail.com (A.G.); amerikanou@windowslive.com (C.A.); ngalanis@gmail.com (N.G.)

* Correspondence: akaliora@hua.gr or andrianakaliora@gmail.com; Tel.: +30-210-9549226; Fax: +30-210-9577050

Received: 19 June 2020; Accepted: 27 July 2020; Published: 30 July 2020

Abstract: Monoterpenes are bioactive compounds, however studies on their metabolic fate in humans are scarce. The present work aimed to identify and quantify the bioactive monoterpenes myrcene, α- and β-pinene of the Mediterranean product Mastiha Oil, in human plasma after acute consumption of a single dose. This was an open-label, single-arm acute study. After overnight fasting, healthy males were administered with Mastiha Oil. Blood samples were collected on different time-points before and after consumption. A novel GC-MS-MS application was performed to detect and quantify terpenes in MO and in plasma. Serum lipid resistance to oxidation was also determined. Alpha-Pinene, β-pinene and myrcene were identified and quantified in plasma. Alpha-pinene concentration significantly increased after 0.5 h of Mastiha Oil consumption, remaining significantly increased at 1 h, 2 h, 4 h, 6 h and 24 h. Beta-pinene and myrcene followed similar patterns. The increase in serum lipid resistance to oxidation was significant at 1 h, reached its peak at 2 h and remained significant until 4 h. Conclusively, α-pinene, β-pinene and myrcene that are present in Mastiha Oil are absorbed by humans. (ClinicalTrials.gov Identifier: NCT04290312).

Keywords: monoterpenes; mastiha oil; absorption

1. Introduction

Natural and plant-derived products or additives contain large amounts of non-nutrient phytochemicals. There is abundant research upon their effects on health, with several studies revealing beneficial properties on the prevention of chronic diseases, mainly through the activation of the immune system [1]. Nevertheless, evidence on the fraction of administered compounds that can pass into the plasma and body tissues without changing their structure, is yet very limited and often zero. This step is necessary to understand bioactivity and to prove efficacy.

Mastiha Oil (MO) is extracted from the resin of Mastiha, the dried resinous exudate from stems and branches of Pistacia lentiscus. Mastiha is a concentrated source of monoterpenes (e.g., α-pinene, β-pinene, β-myrcene) [2], triterpenes (e.g., mastihadienonic acid, isomastihadienonic acid) [3], and to a smaller degree of plant sterols, simple phenols and approximately 10% MO [4]. MO is a 100% natural product of the Mediterranean and it is suitable for human consumption as it is processed according to the legal standards. Previous research has shown that a total of 90 compounds have been identified in MO (50% monoterpene hydrocarbons, 20% oxygenated monoterpenes, 25% sesquiterpenes) [5]. Monoterpenes seem to exert favorable health effects by regulating mechanisms of oxidative stress and inflammation [6,7]. Additionally, previous studies have shown numerous valuable properties of these compounds such as antibiotic resistance modulation, antitumor, anticoagulant, antimicrobial, anti-Leishmania, antimalarial and analgesic effects [8–10].

We have previously investigated for the first time the kinetics and bioavailability of Mastiha's main triterpenes in healthy subjects. Additionally, we have shown an antioxidant potential after acute consumption of Mastiha [11]. The aim of the present study was to investigate the absorption of the main monoterpenes of MO in humans for the first time, namely α-pinene (PubChem CID: 6654), myrcene (PubChem CID: 31253) and β-pinene (PubChem CID: 440967). To achieve this we employed a novel GC-MS-MS application, so that we overcome matrix difficulties with the tandem MS technique. Additionally, the effect on human antioxidant capacity was assessed with the serum oxidisabilty assay based on the existing data regarding the antioxidant properties of monoterpenes.

2. Experimental methods

2.1. Materials

All monoterpene standards were purchased from Sigma-Aldrich (St. Louis, MO, USA) and were of high purity (>97%). All organic solvents were at least of LC grade and purchased from Merck KGaA, Darmstadt, Germany. MO was kindly donated by The Chios Gum Mastic Growers Association. MO comes out through distillation process with water and it is 100% natural. MO is resinous liquid, colourless to faintly yellow and has a balsamic, green and rustic odour. The density was 0.83 g/mL (20 °C). Its nutritional analysis is presented in Table S1.

2.2. Ethics

Informed consent was obtained by all eligible subjects before they participated in the study. Their rights to physical and mental integrity, to privacy and to the protection of the data were protected in accordance with the Data Protection Act 1998. The study was conducted in accordance with the Declaration of Helsinki and Tokyo for humans and the Principles of Good Clinical Practice. Harokopio University Ethics Committee (49/29-10-2015) approved the protocol. The trial was registered on ClinicalTrials.gov and was assigned with Identifier NCT04290312.

2.3. Study Design

An open label intervention trial was designed to assess the postprandial plasma levels of the main monoterpenes present in MO and their effect in antioxidant capacity. A kinetic approach was applied. Fifty apparently healthy male adults were screened. Ten volunteers were found eligible to the study according to specific criteria (Table 1). Eligible subjects were informed about the aims, methodology, expected effect and any risk from the Information Leaflet of the study, before signing an Informed Consent. The investigator explained that participation was voluntary and participants received a copy of the signed Informed Consent.

Table 1. Inclusion and exclusion criteria.

Inclusion Criteria	Exclusion Criteria
Sex: Male	Obesity
Age: 20–40 years old	Alcohol or drug abuse
	Medication, vitamins and inorganic supplements
	Vegan and macrobiotic diet before and during the trial
	Gastrointestinal diseases (i.e., atrophic gastritis, Inflammatory Bowel Disease, peptic ulcer, gastrointestinal cancer)

2.4. Postprandial Study Protocol

2.4.1. Baseline Assessment

A medical and dietary assessment took place after enrollment. A blood sample was collected for complete blood count, fasting glucose, total cholesterol, triglycerides, low-density lipoprotein (LDL), high-density lipoprotein (HDL), urea, creatinine, serum glutamic oxaloacetic transaminase (SGOT), serum glutamic pyruvic transaminase (SGPT), alkaline phosphatase (ALP), serum gamma glutamyltransferase (γ-GT), lactate dehydrogenase (LDH) and total bilirubin measurement on an automatic biochemical analyzer (Cobas 8000 analyser, Roche Diagnostics GmbH). In addition, anthropometric indices were measured twice. Body weight was measured with a weighting scale (Seca) to the nearest 0.1 kg, while height was measured with a standard stadiometer (Tanita) to the nearest millimeter. Body Mass Index was calculated. In addition, body composition (fat mass, muscle mass, bone mass, total body water) was assessed with Bioelectric Impedance Analysis (Tanita, SA165 A-0950U-3).

2.4.2. Intervention

After overnight fasting, a plastic cannula was inserted in an arm vein of the volunteers in order discomfort during consecutive blood sampling to be minimized. A blood sample was obtained on time-point 0 h and after that the subjects consumed 1 mL of MO. This dose was selected based upon the study of Papada et al. [11] where healthy volunteers were administered with 10 g of Mastiha (containing ~10% MO). Afterwards, blood samples were collected on time-points 0.5 h, 1 h, 2 h, 4 h, 6 h and 24 h after MO consumption (Figure 1) and were centrifuged at 3000 rpm for 10 min at 4 °C for plasma and serum isolation. In this single-arm study all volunteers were assigned to the intake of monoterpenes present in MO. Measurements on time-point 0 h served as control, as these are considered known constants in absorption studies. All samples were stored at −80 °C until further analysis.

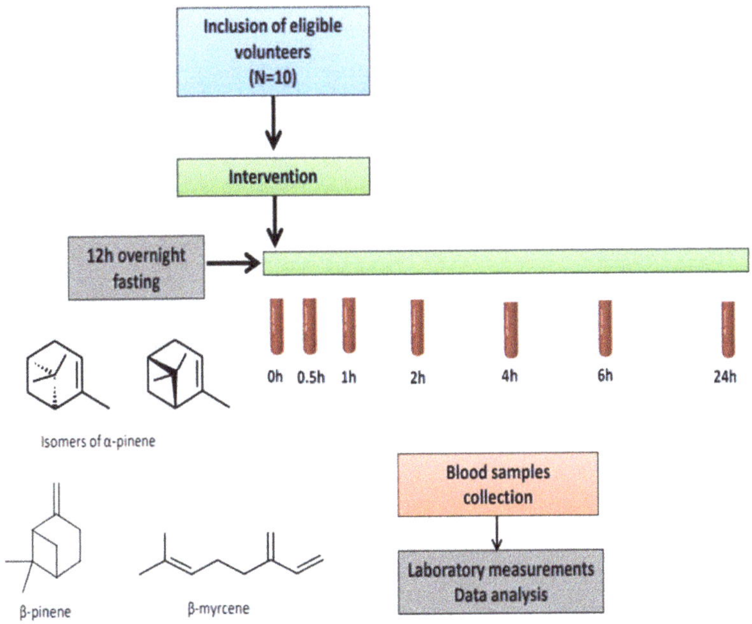

Figure 1. Study flowchart.

2.5. *Analytical Techniques and Assays*

2.5.1. Detection and Quantification of Monoterpenes in Plasma Samples

A Thermo Ultra GC-Thermo XLS Quantum GC-MS-MS Triple Quad with a DB-5 ms 30 × 0.25 × 0.25 capillary column (Merck KGaA, Darmstadt, Germany) was used. The sample was injected using a Programmed Temperature Vaporizing (PTV) injector that allows concentration of samples, using an automated sampler. Injection volume was 1 μL.

To be able to determine low ppb levels, a GC-MS-MS technique was performed. Signal to noise improves radically compared to traditional GC-MS methods. Full scan MS-MS was first employed in analytical standards which run individually to determine the parent and daughter ions and retention times were validated with the use of analytical standards which run individually.

Gas Chromatograph conditions:

Initial temperature was held at 50 °C for 2 min. Then temperature was ramped to 60 °C with a rate of 5 °C/min. The temperature reached 270 °C at a rate of 7 °C/min. Injector temperature was set at 250 °C. The splitless injected volume was 1 μL. At 1.2 min the splitless valve switched.

The MS-MS conditions were as follows: Parent Ion was m/z 93 and daughter ions were 91 and 77 for all compounds. Collision energy was set at 8 V and collision gas pressure was 1.5 mTorr. MS-MS scan time was set at 0.15 s. The peak width was maintained at 0.7 amu.

For sample preparation, a 1:1 mixture of ethyl-acetate:hexane was used and equal volumes of blood plasma and extractant were put in centrifuge tubes and vortexed for 20 min. The samples were then centrifuged and filtrated before injection.

To be able to determine low ppb levels, a GC-MS-MS technique was performed, which, compared to traditional GC-MS methods, improves signal to noise values and thus sensitivity in the identification of compounds. Full scan MS-MS was first employed in analytical standards which run individually to determine the parent and daughter ions of the three terpenes and their retention times. Identification of chromatographic peaks was accomplished by using the parent ion and most abundant daughter ion of each compound and comparing the retention times with those of reference standards. For quantification, blank plasma samples (collected prior to MO administration) were spiked with standard monoterpene solutions (1–2000 μg/L), and were treated identically to the samples, since the standard addition method used eliminates the matrix effect. The standard monoterpene solutions were prepared in ethyl acetate in the range 2–2000 μg/L by diluting stock solutions (1 mg/mL). In all preliminary tests the average recovery for the monoterpenes studied was >80% and the detection limit was 0.02 μg/L.

2.5.2. Kinetics of Monoterpenes in Plasma

The maximum concentrations of monoterpenes (Cmax) and the time intervals (Tmax) were obtained from each participant's plasma concentration–time curve. The area under the plasma concentration–time curve (AUC) was analysed using the Linear Trapezoidal method.

2.5.3. Antioxidant Capacity

The susceptibility of serum lipoproteins to oxidation induced by copper sulphate was estimated as previously described [12]. The produced conjugated dienic hydroperoxides were measured in an Elisa reader (Biotek PowerWave XS2). The increase in absorbance (245 nm) was plotted against time. The lag-time preceding oxidation was expressed in seconds (tLAG) according to Esterbauer and Jurgens [13]. All experiments were conducted at least in duplicate.

2.6. *Statistics*

Data were analysed with the Statistical Package for the Social Sciences (SPSS 21.0, SPSS Inc., Chicago, IL, USA). Non-parametric Friedman's test was applied to evaluate differences in terpenes concentration and total serum oxidizability between the time-points. Level of statistical significance was set at $p < 0.05$. All data are presented as mean values ± standard error of mean (SEM). The G*Power

3 software (HHU, Dusseldorf-Germany, 2007) program was used to calculate the sufficient sample size with an αlpha-value of 5% and a power of 0.8. The sufficient sample size obtained was eight participants.

3. Results

All subjects completed the intervention. No adverse effects of MO intake were reported. Table 2 presents anthropometry and biochemical profile at baseline. Biochemical indices of hepatic and renal function were quantified to assess the safety of the dose administered (Table S2). As depicted in Table 2 and Table S2, all the parameters were within the normal range.

Table 2. Baseline characteristics.

Anthropometric	
Age (year)	25.8 ± 2.4
Height (cm)	178.0 ± 1.7
Weight (kg)	85.2 ±3.7
BMI (kg/m^2)	26.9 ± 1.2
Body fat (%)	19.9 ± 2.9
Total Body Water (%)	57.1 ± 2.0
Muscle mass (kg)	64.2 ± 1.4
Bone mass (kg)	3.3 ± 0.0
Biochemical	
Red Blood Cells (/μL)	5,256,250.0 ± 111,624.9
White Blood Cells (/μL)	6923.8 ± 646.6
Hemoglobin (g/dL)	16.1 ± 0.4
Hematocrit (g/dL)	46.1 ± 1.0
Mean Corpuscular Volume (fl)	87.8 ± 1.1
Mean Corpuscular Hemoglobin Concentration (g/dL)	35.0 ± 0.3
Mean Corpuscular Hemoglobin (pg/RBC)	30.7 ± 0.3
Glucose (mg/dL)	89.5 ± 3.4
Total Cholesterol (mg/dL)	182.4 ± 11.6
Triglycerides (mg/dL)	86.7 ± 16.4
High-Density Lipoprotein (mg/dL)	45.3 ± 2.5
Low-Density Lipoprotein (mg/dL)	111.7 ± 7.4
High Risk Rate	4.0 ± 0.4

Values are Mean ± Standard Error of Mean.

3.1. Profiling of MO

The total ion chromatogram (TIC) of extract from MO is presented in Figure S1. The main peaks in MO are respective to α-pinene (82.16%), myrcene (8.53%), and β-pinene (2.41%). Other phytochemicals detected in MO included linalool (0.84%), limonene (0.83%), camphene (0.64%), caryophyllene (0.46%), caryophyllene oxide (0.21%) and verbenone (0.20%).

3.2. GC-MS-MS Analysis of Plasma Samples

The major terpenes of MO, myrcene, α-pinene and β-pinene were detected in plasma samples applying in-house GC-MS-MS based method. A typical chromatogram of a plasma sample of a volunteer on time-point 4 h is depicted in Figure S2. The targeted monoterpenes could be detected already 0.5 h after MO consumption. Myrcene concentration at 0.5 h (774.8 ± 64.0 μg/L) increased significantly compared with baseline ($p = 0.012$), reached its peak at 2 h (905.6 ± 44.6 μg/L) and followed a decreasing trend until 24 h, when its concentration (648.2 ± 60.3 μg/L) was not significantly increased compared with baseline ($p = 0.128$). Alpha-pinene level followed a similar pattern but reached the peak at 4 h (839.1 ± 188.8 μg/L) with this concentration being significantly higher compared with baseline ($p = 0.008$). The level of α-pinene on timepoint 24 h (80.6 ± 7.3 μg/L) remained significantly increased compared with baseline ($p = 0.028$). The same pathway was followed for β-pinene. The peak concentration was observed at 4 h (16.7 ± 3.5 μg/L) and was significantly increased compared with baseline ($p = 0.008$). At 24 h the concentration (2.1 ± 0.3 μg/L) remained increased compared with baseline ($p = 0.046$). The mean plasma concentration-time curves are illustrated in Figure 2.

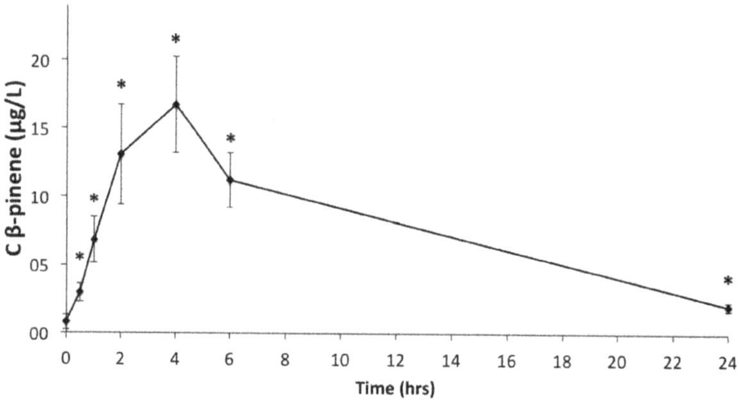

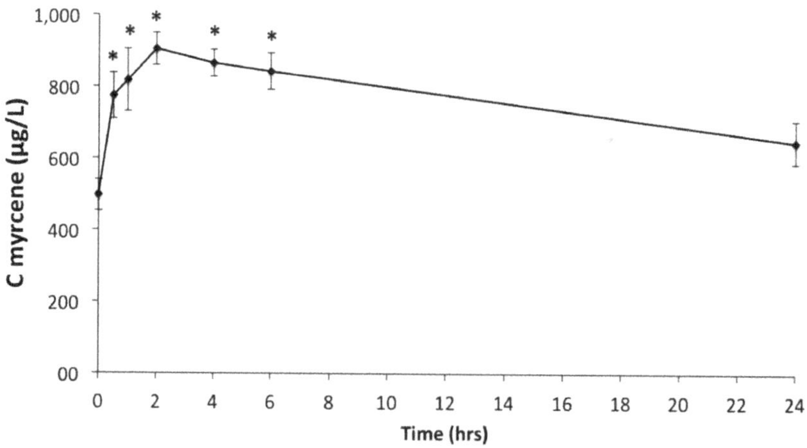

Figure 2. *Cont.*

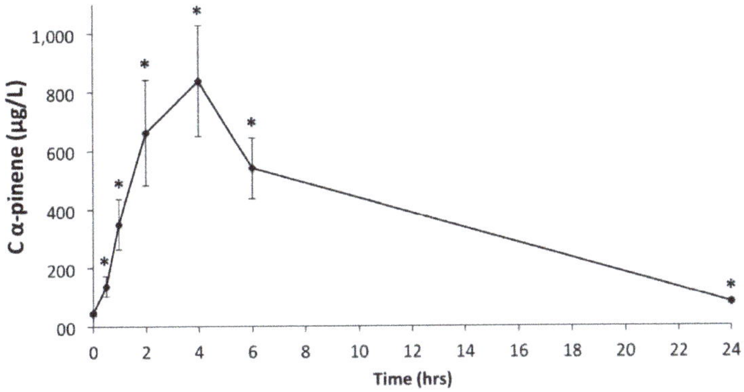

Figure 2. Plasma concentration-time curves for β-pinene, myrcene and a-pinene. Values are presented as mean ± standard error of mean. * p <0.05.

Table 3 depicts the plasma kinetic parameters, namely Cmax, Tmax and AUC. The highest Cmax was observed for myrcene (966.6 ± 89.7 μg/L), while the lowest Cmax was observed for β-pinene (18.0 ± 10.7 μg/L). The highest Tmax was 3.8 ± 1.2 h for α-pinene and the lowest was 2.2 ± 1.7 h for myrcene. The highest AUC was observed for myrcene and the lowest for β-pinene.

Table 3. Kinetics of Myrcene, α-Pinene and β-Pinene in plasma.

Terpenes	C_{max} (μg/L)	T_{max} (h)	Area Under Curve (μg·h/L)
Myrcene	966.6 ± 89.7	2.2 ± 1.7	15318.0 ± 7313.3
α-Pinene	914.8 ± 551.2	3.8 ± 1.2	7865.2 ± 5547.3
β-Pinene	18.0 ± 10.7	3.6 ± 0.9	164.0 ± 110.8

Values are Mean ± Standard Error of Mean.

3.3. *Antioxidant Capacity*

The antioxidant capacity is given as lag time (tLAG) in seconds, and particularly as the difference (ΔT) of tLAG of each time point from tLAG 0 h. Antioxidant capacity tended to increase since time point 0.5 h. This rise was significant on 1 h interval (653.2 ± 111.8 s), reached a peak on 2 h interval (1538.5 ± 327.2 s) and remained statistically significant until 4 h post-ingestion (660.1 ± 129.3 s) ($p < 0.05$) (Figure 3).

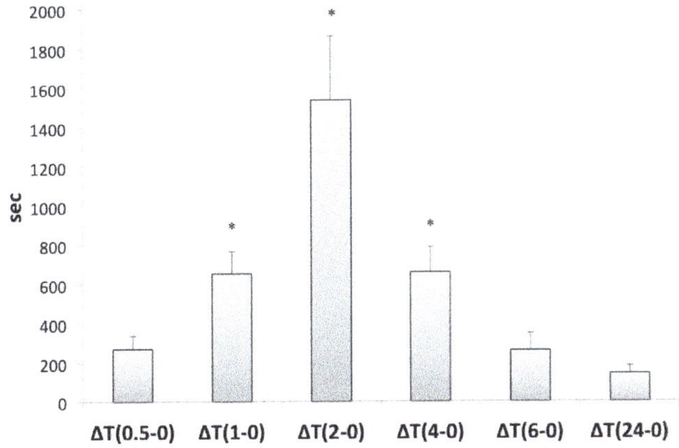

Figure 3. Antioxidant capacity was estimated applying the copper sulphate oxidation assay. * $p < 0.05$.

4. Discussion

Herein, we aimed at assessing whether humans absorb monoterpenes that are abundant in MO. Previous experiments on rats set the oral lethal dose—LD50—of MO at a dose of 5 g/kg body weight [14]. Our results show for the first time that MO consumption at the dose of 1 mL is safe, since there were no adverse effects and the parameters of renal and hepatic function were within the normal range after MO consumption. This finding is crucial, since natural products like MO could serve as potent weapons against pathological conditions related with oxidative stress and inflammation. This fact coincides with the postprandial kinetic study of Papada et al. with Mastiha at a dose of 10 g (containing approx. 1 mL MO) [11].

According to our knowledge this is the first study showing that the major terpenes of MO are bioavailable in human plasma using a method with a significant methodological advantage. To be able to determine low ppb levels in a complicated matrix like blood plasma, a GC-MS-MS technique was performed. Signal to noise improves significantly compared to traditional GC-MS methods.

Our results showed that myrcene, α-pinene and β-pinene are detected in plasma already after 0.5 h reaching their peaks at 2 h to 4 h post-ingestion. These time-points were slightly different; a study on rats showed that α-pinene was also bioavailable by 0.17 h reaching its peak concentration in plasma 2.5 h after oral administration [15]. However, it is important to mention that human and animal metabolism share not only common but also different metabolic characteristics that do not allow us direct comparisons and generalizations. As regards to our previous study in humans investigating the absorption of triterpenes, we have shown similar patterns of kinetics. The main triterpenes of Mastiha were detected 0.5 h after consumption and reached their peak concentration between 2 h and 4 h. These patterns coincided with the present study but cautious interpretation is necessary, since Mastiha has a different matrix compared to MO and additionally triterpenes are much more complex compounds than monoterpenes [11]. The fact that monoterpenes of MO reach the plasma unaltered could partially explain the beneficial properties for human health that have been attributed to MO since antiquity.

Herein the retention time of b-pinene was found lower to the respective of myrcene. This result is in compliance with the study of Thao et al. [16] using a similar DB5 column. Overall, when the natural product was profiled, more terpenes were detected compared to the blood samples. These results are of no surprise since metabolism of terpenes is quite complex. Different pH conditions, mechanical and enzymatic activities, and also transformations into generally more water-soluble and more readily excreted in the urine compounds take place. These changes are observed primarily in the liver, gastrointestinal tissue, lungs, kidneys, brain and blood [17].

As regards oxidative stress, we evaluated serum resistance to oxidation, which was significantly increased on time point 1 h, reached a peak on time point 2 h and remained statistically significant until 4 h post-ingestion. In our previous study on the absorption of triterpenes that are abundant in Mastiha, serum resistance to oxidation increased significantly at 4 h, reached its peak at 6 h and remained significantly at 24 h. This effect was associated with the availability of the detected triterpenes in plasma. Most studies evaluating the absorption of phytochemicals have reported a peak in serum antioxidant capacity 1–2 h after ingestion, a finding that comes into agreement with our results. A postprandial study of Kanellos and coworkers assessed the absorption and plasma kinetics of phytochemicals present in raisins and their effect on serum oxidation resistance in healthy humans [18]. Serum resistance to oxidation and plasma total phenolics reached their peak 1 h after raisin consumption. Despite the similar patterns revealed, the different matrices of the foods been studied should not allow us to make direct comparisons.

It is also important to mention that the terpenes detected and quantified in plasma could affect directly oxidative stress, but also endogenous antioxidants may be implicated in the increase of resistance to oxidation. However, we could hypothesize that the differences in antioxidant capacity could be attributed to terpenes that have proven antioxidant properties [19]. A study on the essential oil from black pepper that contains myrcene, α-pinene and β-pinene pointed out antioxidant properties in vitro [20]. More specifically, the essential oil scavenged diphenyl-2 picrylhydrazyl (DPPH), nitric oxide (NO), 2,2′-azino-bis(3-ethylbenzthiazoline-6-sulphonic acid)ABTS and chelated Fe^{2+}. Although further research is necessary to investigate the direct effects of monoterpenes on antioxidant defense mechanisms in humans, our results could point towards this direction.

Our protocol has significant strengths. Gender differences, the menopause and menstrual cycle with hormonal fluctuations in females seem to affect the absorption of phytochemicals [21,22]. Thus, the recruitment of males is one of our protocol's advantages so that the kinetics of the monoterpenes is not affected by those factors. Additionally, strict adherence to the inclusion and exclusion criteria during recruitment of participants led to a homogenous sample. An additional strength is the application of a GC-MS-MS technique not previously described in the monoterpene detection in human plasma. This application allows for the identification and quantification of these aromatic phytochemicals even at low detection levels without the matrix interferences.

5. Conclusions

This is the first study showing that myrcene, alpha- and beta pinenes abundant in MO and several natural products are bioavailable in plasma already 30 min after consumption of a single dose. Additionally, they may contribute to antioxidant defense since serum resistance to oxidation increased 1 h after their administration. Monoterpenes exhibit several beneficial effects and our findings could contribute to understanding their usage and applications on human health. Proving that monoterpenes in MO are bioavailable in human plasma is the first significant step towards the usage of foods and natural products for medicinal purposes. Further research is necessary to characterize the kinetics of these monoterpenes in human metabolism.

Supplementary Materials: The following are available online at http://www.mdpi.com/2304-8158/9/8/1019/s1, Table S1: Nutritional analysis of MO, Table S2: Biochemical parameters of hepatic and renal function, Figure S1: The total ion chromatogram (TIC) of extract from MO, Figure S2: Chromatogram of a plasma sample on time point 4 h.

Author Contributions: Conceptualization-supervision-review and editing, A.C.K.; investigation, E.P., A.G., C.A. and N.G.; writing—original draft preparation, E.P., A.G.; All authors have read and agreed to the published version of the manuscript.

Funding: This research received no external funding.

Acknowledgments: We are thankful to the participants who volunteered in our study. Also, we are grateful to The Chios Gum Mastic Growers Association for their support and for supplying MO.

Conflicts of Interest: The authors declare no conflict of interest.

References

1. Diederich, M. Natural products target the hallmarks of chronic diseases. *Biochem. Pharmacol.* **2020**, *173*, 113828. [CrossRef] [PubMed]
2. Paraschos, S.; Magiatis, P.; Mitakou, S.; Petraki, K.; Kalliaropoulos, A.; Maragkoudakis, P.; Mentis, A.; Sgouras, D.; Skaltsounis, A.-L. In Vitro and In Vivo Activities of Chios Mastic Gum Extracts and Constituents against Helicobacter pylori. *Antimicrob. Agents Chemother.* **2006**, *51*, 551–559. [CrossRef] [PubMed]
3. Assimopoulou, A.N.; Papageorgiou, V.P. GC-MS analysis of penta- and tetra-cyclic triterpenes from resins of Pistacia species. Part I. Pistacia lentiscus var. Chia. *Biomed. Chromatogr.* **2005**, *19*, 285–311. [CrossRef] [PubMed]
4. Kaliora, A.C.; Mylona, A.; Chiou, A.; Petsios, D.G.; Andrikopoulos, N.K. Detection and Identification of Simple Phenolics in Pistacia lentiscus Resin. *J. Liq. Chromatogr. Relat. Technol.* **2004**, *27*, 289–300. [CrossRef]
5. Boelens, M.H.; Jiménez, R. Chemical composition of the essential oils from the gum and from various parts of Pistacia lentiscus l. (mastic gum tree). *Flavour Fragr. J.* **1991**, *6*, 271–275. [CrossRef]
6. Subramaniyan, S.D.; AshokKumar, N. Citral, A Monoterpene Protect Against High Glucose Induced Oxidative Injury in HepG2 Cell In Vitro-An Experimental Study. *J. Clin. Diagn. Res.* **2017**, *11*, BC10–BC15. [CrossRef]

7. Kodikonda, M.; Naik, P.R. Modulatory effect of garcinol in streptozotocin-induced diabetic Wistar rats. *Arch. Physiol. Biochem.* **2017**, *123*, 322–329. [CrossRef]
8. Salehi, B.; Upadhyay, S.; Orhan, I.E.; Jugran, A.K.; Jayaweera, S.L.D.; Dias, D.A.; Sharopov, F.; Taheri, Y.; Martins, N.; Baghalpour, N.; et al. Therapeutic Potential of α- and β-Pinene: A Miracle Gift of Nature. *Biomolecules* **2019**, *9*, 738. [CrossRef]
9. Sieniawska, E.; Swatko-Ossor, M.; Sawicki, R.; Skalicka-Woźniak, K.; Ginalska, G. Natural Terpenes Influence the Activity of Antibiotics against Isolated Mycobacterium tuberculosis. *Med. Princ. Prac.* **2016**, *26*, 108–112. [CrossRef]
10. Matsuo, A.L.; Figueiredo, C.R.; Arruda, D.C.; Pereira, F.V.; Scutti, J.A.B.; Massaoka, M.H.; Travassos, L.R.; Sartorelli, P.; Lago, J.H.G. α-Pinene isolated from Schinus terebinthifolius Raddi (Anacardiaceae) induces apoptosis and confers antimetastatic protection in a melanoma model. *Biochem. Biophys. Res. Commun.* **2011**, *411*, 449–454. [CrossRef]
11. Papada, E.; Gioxari, A.; Brieudes, V.; Amerikanou, C.; Halabalaki, M.; Skaltsounis, L.; Smyrnioudis, I.; Kaliora, A. Bioavailability of Terpenes and Postprandial Effect on Human Antioxidant Potential. An Open-Label Study in Healthy Subjects. *Mol. Nutr. Food Res.* **2017**, *62*, 1700751. [CrossRef] [PubMed]
12. Aurrekoetxea, I.; Ruiz-Sanz, J.I.; Del Agua, A.R.; Navarro, R.; Hernandez, M.L.; Matorras, R.; Prieto, B.; Ruiz-Larrea, M.B. Serum oxidizability and antioxidant status in patients undergoing in vitro fertilization. *Fertil. Steril.* **2010**, *94*, 1279–1286. [CrossRef] [PubMed]
13. Esterbauer, H.; Jiirgens, G. Mechanistic and genetic aspects of susceptibility of LDL to oxidation. *Curr. Opin. Lipidol.* **1993**, *4*, 114–124. [CrossRef]
14. Lewis, C.A.; Palanker, A.L. *Acute Toxicity Studies in Rats and Rabbits*; Unpublished report; Research Institute of Fragrance Materials: Englewood Cliffs, NJ, USA, 1979.
15. Li, W.; Hong, B.; Li, Z.; Li, Q.; Bi, K. GC-MS method for determination and pharmacokinetic study of seven volatile constituents in rat plasma after oral administration of the essential oil of Rhizoma Curcumae. *J. Pharm. Biomed. Anal.* **2018**, *149*, 577–585. [CrossRef] [PubMed]
16. Thao, N.T.; Kashiwagi, T.; Sawamura, M. Characterization by GC-MS of Vietnamese Citrus Species and Hybrids Based on the Isotope Ratio of Monoterpene Hydrocarbons. *Biosci. Biotechnol. Biochem.* **2007**, *71*, 2155–2161. [CrossRef]
17. Furtado, N.A.J.C.; Pirson, L.; Edelberg, H.; Miranda, L.M.; Loira-Pastoriza, C.; Préat, V.; Larondelle, Y.; Andre, C.M. Pentacyclic Triterpene Bioavailability: An Overview of In Vitro and In Vivo Studies. *Molecules* **2017**, *22*, 400. [CrossRef]
18. Kanellos, P.T.; Kaliora, A.; Gioxari, A.; Christopoulou, G.O.; Kalogeropoulos, N.; Karathanos, V.T. Absorption and Bioavailability of Antioxidant Phytochemicals and Increase of Serum Oxidation Resistance in Healthy Subjects Following Supplementation with Raisins. *Plant Foods Hum. Nutr.* **2013**, *68*, 411–415. [CrossRef]
19. González-Burgos, E.; Gómez-Serranillos, M.P. Terpene compounds in nature: A review of their potential antioxidant activity. *Curr. Med. Chem.* **2012**, *19*, 5319–5341. [CrossRef]
20. Oboh, G.; Ademosun, A.; Odubanjo, O.V.; Akinbola, I.A. Antioxidative Properties and Inhibition of Key Enzymes Relevant to Type-2 Diabetes and Hypertension by Essential Oils from Black Pepper. *Adv. Pharmacol. Sci.* **2013**, 926047. [CrossRef]
21. Domínguez-Perles, R.; Auñón, D.; Ferreres, F.; Gil-Izquierdo, A. Gender differences in plasma and urine metabolites from Sprague-Dawley rats after oral administration of normal and high doses of hydroxytyrosol, hydroxytyrosol acetate, and DOPAC. *Eur. J. Nutr.* **2015**, *56*, 215–224. [CrossRef]
22. García-Villalba, R.; Larrosa, M.; Possemiers, S.; Tomás-Barberán, F.A.; Espín, J.C. Bioavailability of phenolics from an oleuropein-rich olive (Olea europaea) leaf extract and its acute effect on plasma antioxidant status: Comparison between pre- and postmenopausal women. *Eur. J. Nutr.* **2013**, *53*, 1015–1027. [CrossRef] [PubMed]

foods MDPI

Article

Yeast Particle Encapsulation of Scaffolded Terpene Compounds for Controlled Terpene Release

Ernesto R. Soto [1], Florentina Rus [1], Hanchen Li [1], Carli Garceau [1], Jeffrey Chicca [1], Mostafa Elfawal [1], David Gazzola [1], Martin K. Nielsen [2], Joseph F. Urban, Jr. [3,4], Raffi V. Aroian [1] and Gary R. Ostroff [1,*]

1 Program in Molecular Medicine, University of Massachusetts Medical School, Worcester, MA 01605, USA; ernesto.soto-villatoro@umassmed.edu (E.R.S.); florentina.rus@umassmed.edu (F.R.); hanchen.li@umassmed.edu (H.L.); carli.garceau@umassmed.edu (C.G.); jeffrey.chicca@umassmed.edu (J.C.); mostafa.elfawal@umassmed.edu (M.E.); davidmgazzola@gmail.com (D.G.); raffi.aroian@umassmed.edu (R.V.A.)

2 M.H. Gluck Equine Research Center, Department of Veterinary Science, University of Kentucky, Lexington, KY 40546, USA; martin.nielsen@uky.edu

3 Beltsville Human Nutrition Research Center, Diet, Genomics and Immunology Laboratory, United States Department of Agriculture, Agricultural Research Service, Beltsville, MD 20705, USA; joe.urban@usda.gov

4 Beltsville Agricultural Research Center, Animal and Parasitic Diseases Laboratory, United States Department of Agriculture, Agricultural Research Service, Beltsville, MD 20705, USA

* Correspondence: gary.ostroff@umassmed.edu; Tel.:+1-508-856-1930

Abstract: Terpenes are naturally occurring compounds produced by plants that are of great commercial interest in the food, agricultural, cosmetic, and pharmaceutical industries due to their broad spectra of antibacterial, antifungal, anthelmintic, membrane permeation enhancement, and antioxidant biological activities. Applications of terpenes are often limited by their volatility and the need for surfactants or alcohols to produce stable, soluble (non-precipitated) products. Yeast particles (YPs) are hollow, porous microspheres that have been used for the encapsulation of terpenes (YP terpenes) by passive diffusion of terpenes through the porous YP cell walls. We here report the development of a second generation YP encapsulated terpene technology that incorporates the stimuli-responsive control of terpene release using biodegradable pro-terpene compounds (YP pro-terpenes). YP terpenes and YP pro-terpenes were both produced, in which high levels of carvacrol, eugenol, thymol and geraniol were encapsulated. The YP pro-terpenes show higher encapsulation stability than YP terpenes due to pro-terpenes being non-volatile solids at room temperature and stable in suspensions at neutral pH. YP pro-terpenes and YP terpenes were evaluated for biological activity in antibacterial, antifungal and anthelmintic assays. The YP pro-terpenes retained the full biological activity of the parent terpene compound.

Keywords: terpenes; yeast particles; antimicrobial; anthelmintic; prodrug

Citation: Soto, E.R.; Rus, F.; Li, H.; Garceau, C.; Chicca, J.; Elfawal, M.; Gazzola, D.; Nielsen, M.K.; Urban, J.F., Jr.; Aroian, R.V.; et al. Yeast Particle Encapsulation of Scaffolded Terpene Compounds for Controlled Terpene Release. *Foods* **2021**, *10*, 1207. https://doi.org/10.3390/foods10061207

Academic Editors: Ian Southwell and Oscar Núñez

Received: 16 April 2021
Accepted: 20 May 2021
Published: 27 May 2021

1. Introduction

Terpenes are a large class of naturally occurring organic compounds that constitute primary components of essential oils obtained from plants. The term terpene is commonly used to include basic terpenes or hydrocarbons consisting of isoprene repeating units and terpenoids or modified terpenes containing additional functional groups (usually oxygen-containing groups). Terpenes have long been recognized for a wide range of functional properties (e.g., anthelmintic, antifungal, antibacterial, and antioxidant properties). Terpenes are used as excipients due to their permeation enhancer properties, or as bioactive compounds in pharmaceutical products, as fragrances, permeation enhancers and antioxidants in cosmetics, for their broad range of potential applications against pathogens in agricultural products, and can be used as additives in food packaging to prevent bacterial spoilage and oxidation [1–3].

The use of terpenes can present challenges primarily associated with (1) the chemical instability of some isolated terpenes when exposed to air, heat, light and moisture; (2) poor water solubility, and (3) high volatility. It is usually necessary to produce terpene-based products containing high levels of surfactants or alcohols. Applications of terpenes in certain products such as food preservatives poses challenges due to safety limits and marked organoleptic effects. Encapsulation of terpenes in nano or microstructured systems has been employed as a common approach to develop terpene formulations with improved chemical stability, shelf-life, additionally opening up the possibility to control terpene release [4–8].

We have developed methods using yeast particles (YPs) to efficiently encapsulate high levels of terpenes. YPs are 3–5 µm hollow and porous microspheres, a byproduct of the food grade Baker's yeast (*Saccharomyces cerevisae*) extract manufacturing process. We have used YPs for the encapsulation of a broad range of molecules for drug delivery and agricultural applications [9–15]. Our first-generation approach to load terpenes inside the hydrophobic cavity of YPs is based on passive diffusion of terpenes through the porous microsphere cell walls without the need for surfactants or alcohols. Short-term sustained terpene release from YPs is the reverse process and is a function of terpene water solubility. This approach has been successfully implemented to develop and commercialize a YP-terpene based fungicide and nematicide for agricultural applications [16–21]. More recently, we have identified YP terpenes with broad-acting anthelmintic activity [22], which could lead to the development of formulations for oral terpene delivery for the treatment of gastrointestinal worm parasites and other infectious agents.

To develop better controlled terpene release, as well as to avoid limitations associated with terpene volatility and stability, we have developed a second generation YP terpene encapsulation approach using non-volatile, biodegradable pro-terpene compounds. The pro-terpene compounds are (1) solids at room temperature with a high melting point to avoid terpene loss due to the high volatility of terpenes, (2) water insoluble to avoid premature release from aqueous-based YP compositions, (3) stable at neutral pH, and (4) susceptible to chemical (pH) or enzymatic hydrolysis of a biodegradable linker providing for controlled release. This approach was first demonstrated for the YP encapsulation of a pro-carvacrol compound [23]. The YP pro-carvacrol was shown to have similar in vitro antibacterial properties as YP carvacrol, but with the additional benefits of improved stability and controlled, sustained carvacrol release from YPs.

In this article, we report the use of the YP pro-terpene encapsulation approach of four terpenes: carvacrol, eugenol, thymol, and geraniol and expanded testing against a wider range of applications. The YP terpene and YP pro-terpene compositions were characterized for controlled terpene release and screened in vitro for antibacterial, antifungal, and anthelmintic activity in model systems. The results show the potential broad range of applications of YP terpenes and YP pro-terpenes, and the advantage of YP pro-terpene encapsulation to improve the formulation stability and to control terpene release.

2. Materials and Methods

2.1. Materials

Yeast Particles (YPs) were purchased from Biorigin (Louisville, KY, USA). Terpenes (carvacrol, eugenol, thymol, and geraniol) were procured from Penta Manufacturing (Livingston, NJ, USA). All reagents and solvents for synthesis and HPLC analysis were obtained from Fisher Scientific (Waltham, MA, USA) or Sigma Aldrich (St. Louis, MO, USA). Lysogeny broth (LB) was purchased from Sigma Aldrich and yeast peptone dextrone (YPD) was prepared from Difco™ yeast extract, Difco™ Bacto peptone and dextrose (all materials obtained from Fisher Scientific) at a composition of 1% yeast extract, 2% peptone and 2% dextrose. Reagents for worm culture medium were purchased from Gibco (Gaithersburg, MO, USA).

2.2. *Methods*

2.2.1. Synthesis of Pro-Terpene Compounds

Pro-terpenes were synthesized from the reaction of the parent terpene compound and ethylenediaminetetracetic acid (EDTA) dianhydride based on a previously described procedure for phenolic terpenes [23,24]. Briefly, terpene (18 mmol) and triethylamine (TEA, 64 mmol) were dissolved in 75 mL of anhydrous tetrahydrofuran (THF). EDTA dianhydride (9 mmol) was added slowly to the THF solution. The reaction mixture was stirred under nitrogen, at room temperature, overnight. The mixture was diluted in 500 mL of water and concentrated HCl was added to immediately acidify to pH 2. The precipitated product was filtered, washed with water, and dried under vacuum.

Pro-carvacrol–Yield: 71%; off-white powder; ^{1}H-NMR (DMSO-d_6, 500 MHz): δ ppm, 7.2 (d, 2H, Ar-H); 7.04 (d, 2H, Ar-H); 6.9 (s, 2H, Ar-H); 3.9 (s, 4H, CH_2); 3.67 (s, 4H, CH_2); 3.03 (s, 4H, CH_2); 2.89 (m, 2H, CH); 2.05 (s, 6H, CH_3); 1.1 (d, 12H, CH_3)

Pro-eugenol–Yield: 81%; off-white powder; ^{1}H NMR (DMSO-d_6, 500 MHz): δ ppm, 6.9–7.0 (m, 4H, Ar-H); 6.7–6.8 (d, 2H, Ar-H); 6.0 (m, 2H, CH); 5.0–5.1 (m, 4H, CH_2); 3.8 (s, 4H, CH_2); 3.7 (s, 6H, CH_3); 3.5 (s, 4H, CH_2); 3.4 (d, 4H, CH_2); 2.9 (s, 4H, CH_2)

Pro-thymol–Yield: 60%; off-white powder; ^{1}H NMR (DMSO-d_6, 500 MHz): δ ppm, 7.2 (d, 2H, Ar-H), 7.05 (d, 2H, Ar-H); 6.85 (s, 2H, Ar-H); 3.9 (s, 4H, CH_2); 3.6 (s, 4H, CH_2); 2.9 (s, 4H, CH_2); 2.8 (m, 2H, CH); 2.2 (s, 6H, CH_3); 1.1 (d, 12 H, CH_3)

Pro-geraniol–Yield: 27%; off-white powder; ^{1}H NMR (DMSO-d_6, 500 MHz): δ ppm, 5.35 (m, 2H, CH); 5.1 (m, 2H, CH); 4.6 (m, 4H, CH_2); 3.95 (s, 4H, CH_2); 3.6 (s, 4H, CH_2); 3.1 (s, 4H, CH_2); 2.0–2.1 (m, 8H, CH_2); 1.65 (m, 12 H, CH_3); 1.55 (s, 6H, CH_3)

2.2.2. YP Loading of Terpenes (YP Terpene)

Dry YPs were mixed with 0.5 µL water per mg YP. Then, terpene was absorbed into YPs by adding 1 mg terpene per mg YP and incubated at room temperature for 18–24 h for samples containing carvacrol, geraniol or eugenol, and at 65 °C for 18–24 h for samples containing thymol.

2.2.3. YP Loading of Pro-Terpenes (YP Pro-Terpene)

Dry YPs were mixed with 0.5 µL water per mg YP. Then, pro-terpene was absorbed into YPs by swelling the particles with a solution of pro-terpene in DMSO (2.5 µL/mg YP). The samples were incubated at room temperature for 18–24 h to complete the loading. The YP pro-terpene was then lyophilized, and the loading process was repeated until the target concentration of encapsulated pro-terpene was achieved. YP pro-terpene samples contained 1.78–1.85 mg pro-terpene per mg YP to yield 1 mg terpene/mg YP upon pro-terpene hydrolysis.

2.2.4. Characterization of Terpene and Pro-Terpene Loading Efficiency

Samples were stained with Nile red to qualitatively assess loading by the fluorescence microscopy of the encapsulated terpene or pro-terpene Nile red complex. YP terpene and YP pro-terpene samples were suspended in water at a concentration of 10 mg YP/mL. The samples were centrifuged to collect excess liquid (free terpene or pro-terpene). The supernatants and YP pellets were incubated in 0.1 M sodium carbonate buffer (pH 10) for 3 h to hydrolyze pro-terpene and then the samples were diluted with methanol to quantify free (supernatant fraction) and YP encapsulated (pellet fraction) terpene. Terpenes were quantified by HPLC [22] operated with 32 Karat™ software version 7.0 (Beckman Coulter, Inc, Brea, CA, USA), using a Waters Symmetry® C18 column (3.5 µm, 4.6 × 150 mm) with acetonitrile:water 70:30 as mobile phase, flow rate at 1 mL/min, injection volume of 10 µL, and terpene detection at 254 nm. This isocratic HPLC method allows for the detection of single terpene samples with the following retention times: 3.35 min (carvacrol), 2.73 min (eugenol), 3.35 min (geraniol), and 3.42 min (thymol). The quantification of terpenes was done by measuring the peak area and interpolating the concentration using a calibration curve obtained with terpene standards.

2.2.5. Terpene Release from YPs

YP terpene and YP pro-terpene samples were suspended in phosphate buffer saline (PBS, pH 7) at a concentration of 1 mg YP/mL (=1 mg terpene/mL) and incubated at 37 °C. Aliquots were collected at predetermined times, centrifuged and the supernatant was collected to quantify terpene released from the particles by HPLC.

2.2.6. Antimicrobial Activity Assays in Model Bacterial and Fungal Organisms

The antimicrobial activity of YP terpene and YP pro-terpene was evaluated using a modified microplate assay published procedure [25]. Samples of YP terpene and YP pro-terpene were suspended in 100 μL of growth medium (LB was used in antibacterial assays and YPD in antifungal assays) and added to the first row (Row A) of a 96-well plate (all wells in the 96-well plate contain additional 100 μL medium). Serial dilution (1:1) was performed by transferring 100 μL from Row A to Row B, etc., and finally removing 100 μL from Row H. Diluted *Escherichia coli* Top10 (Invitrogen, Carlsbad CA) or *Saccharomyces cerevisae* Cry1 [26] cells (100 μL, 10^6 cells/mL) were added to all wells of the plate. Initial (t = 0) and final (t = 16 h, 37 °C) absorbance readings were taken at 650 nm. The minimum inhibitory concentration (MIC) was determined as the concentration that inhibits bacterial or fungal growth by more than 75%.

2.2.7. Antimicrobial Assay with Fractionated Samples

YP pro-carvacrol samples (10 mg carvacrol/mL) were incubated 18 h at 37 °C in fresh LB and LB collected from *E. coli* culture (spent LB). The samples were centrifuged to collect the supernatant and YP pellet fractions; an aliquot of the supernatant was used to quantify released carvacrol by HPLC. The pellet fractions were suspended in water and an aliquot was used for HPLC analysis. Both supernatant and pellet fractions used for HPLC analysis were incubated in 0.1 M carbonate buffer (pH 10) for 3 h to hydrolyze pro-terpene and then diluted with methanol (final composition of 90% methanol) to solubilize terpene. The remaining supernatant and pellet samples were evaluated for antibacterial activity on *E. coli*, as described above.

2.2.8. Simulated Digestion Assay

YP terpene and YP pro-terpene samples were suspended in simulated gastric fluid (SGF) [27] containing 3.2 mg pepsin/mL at a concentration of 1 mg terpene/mL. The samples were incubated for 1 h at 37 °C, centrifuged and the SGF supernatant was collected. The pellet was suspended in simulated intestinal fluid (SIF) [28] containing 10 mg pancreatin/mL, incubated at 37 °C for 1 h, and centrifuged to collect SIF supernatant. The pellet was suspended again in SIF with fresh pancreatin, and incubated at 37 °C for 1 h. Terpene released from YPs was quantified in all supernatants by HPLC.

2.2.9. Nematode Extract Assay

Ascaris suum 4th stage larvae (350 mg) were isolated from pig intestines between 14–21 days after inoculation [29] and were immersed in PBS (pH 7) or 0.1 M sodium acetate buffer (pH 5) and sonicated with a microtip sonicator probe at maximum power multiple times for 30 s until the worm was disintegrated. Samples were kept on ice during sonication. Total protein in worm extracts was quantified using the bicinchoninic acid protein assay kit (ThermoFisher Scientific, Waltham, MA, USA) with bovine serum albumin control (10.3 ± 1.7 mg protein/mL in worm extract pH 7, 9.2 ± 0.5 mg protein/mL in worm extract pH 5). YP pro-carvacrol samples were suspended in worm extract at a target carvacrol concentration of 20 mg/mL for 24 h at 37 °C. Control YP pro-carvacrol samples were incubated in PBS, acetate buffer, and 0.1 M carbonate buffer pH 10. Samples were diluted 1:1 with methanol, centrifuged, the supernatant was collected and the released carvacrol was quantified by HPLC. Supernatants and YP pro-carvacrol pellets were evaluated for antimicrobial activity on *E. coli*.

2.2.10. Larval Development (Egg-to-Larvae (E2L)) Assay

An E2L assay was used to measure the effects of carvacrol and pro-carvacrol on the development of cyathostomin larvae from eggs to third-stage larvae (L3) [30]. Cyathostomin eggs were collected from the feces of an equine herd [31]. Approximately 60 nematode eggs were added to each well of a 96-well plate using a repeat dispensing pipette. YP samples (10 µL of each working solution) were impregnated into 90 µL of S medium with *E. coli* OP50. Control wells received 10 µL of water only. The plates were incubated for 7 days at 28 °C, larvae were then killed using Lugol's iodine, and the number of fully grown infective L3 were counted in each well. Each concentration-response experiment consisted of triplicate wells.

2.2.11. Adult Worm In Vitro Screening

Ancylostoma ceylanicum worms were maintained in golden Syrian hamsters, as previously described [32]. *Trichuris muris* parasites were maintained in STAT6-/-mice [33]. Adult worms were harvested from infected rodents and washed using prewarmed medium (RPMI 1640 with 25 mM HEPES (pH 7.2) and antimicrobials (100 U/mL penicillin, 100 µg/mL streptomycin, fungizone (10 µg/mL for *A. ceylanicum*, 2.5 µg/mL for *T. muris*)). Worms were manually picked into the wells of the 48-well screening plate (1 worm per well) containing 250 µL RPMI per well. As serum is incompatible with the assays with YP-terpenes, it was left out. YP terpene and YP pro-terpene samples were evaluated at a concentration of 333 µg terpene/mL [22]. Assay plates were incubated at 37 °C and 5% CO_2. Terpene activity was determined by motility of adult worms measured with an in-house assembled Worminator [34]. The Worminator consists of a dark field illuminator, plate holder and video camera placed under the assay plate. Worm motility was recorded using the "WormAssay" software [34], measuring the average motility in each well based on pixel displacement between frames over a given time. Data is expressed as mean motility units and percent inhibition of motility was calculated relative to the mean motility units of control worms (worms incubated with media).

3. Results

3.1. Preparation and Characterization of Yeast Particle Encapsulated Terpenes and Pro-Terpenes

3.1.1. Synthesis of Pro-Terpenes

To improve the stability of YP-terpenes and to better control sustained terpene release, we designed pro-terpene compounds with the following properties: (1) solid at room temperature to prevent terpene loss due to high volatility of terpene compounds, (2) water insoluble to avoid premature terpene release and loss in diluted samples, (3) stable at neutral pH, and (4) susceptible to chemical or enzymatic hydrolysis of a biodegradable linker providing for controlled terpene release.

Pro-terpenes were synthesized via ring-opening transesterification of EDTA dianhydride with the hydroxyl group of terpenes (carvacrol, eugenol, thymol, and geraniol) in the presence of triethylamine (TEA) to yield diacids of terpenes with ester biodegradable bonds, as depicted in Figure 1. The pro-terpenes have high melting points compared to terpenes (three of the terpenes are liquid at room temperature) and are practically insoluble in water (Table 1). The high melting point, poor water solubility, and biodegradable linkers make these pro-terpenes suitable candidates for the development of stable YP pro-terpene formulations with controlled terpene release.

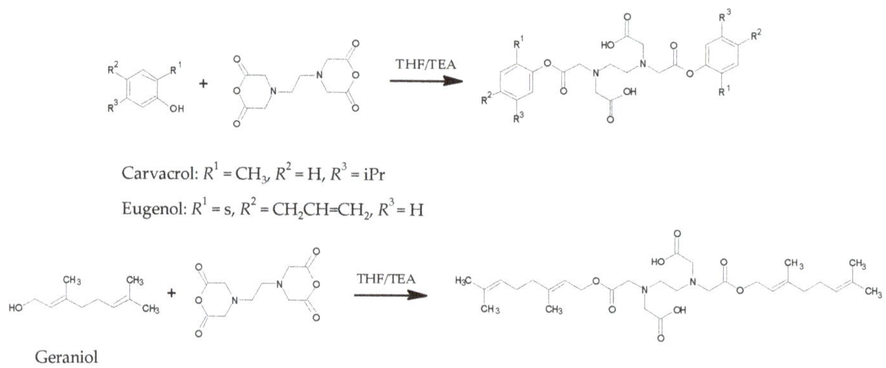

Figure 1. Synthesis of pro-terpenes.

Table 1. Melting point and solubility in water of terpenes and pro-terpenes.

Compound	Melting Point (°C)	Solubility in Water
Carvacrol [1]	3.5	1.25 mg/mL
Pro-carvacrol [2]	156–158	Insoluble
Eugenol [1]	−9.2	1.44 mg/mL
Pro-eugenol [2]	167–169	<20 μg/mL
Thymol [1]	51.5	0.90 mg/mL
Pro-thymol [2]	135–138	Insoluble
Geraniol [1]	−15	0.1 mg/mL
Pro-geraniol [2]	Decomposes above 145	Insoluble

[1] Terpene data from PubChem [35]. [2] Pro-terpene data determined experimentally.

3.1.2. Yeast Particle Encapsulation of Terpenes and Pro-Terpenes

YPs are 3–5 μm, hollow and porous microparticles derived from Baker's yeast. The porous cell wall structure makes these particles excellent absorbent materials, and payloads can be loaded from aqueous and some organic solutions with high payload loading capacity and efficiency in the large hollow cavity of the particles. Our first-generation terpene encapsulation approach is depicted in Figure 2a and is based on the loading of terpenes inside the hydrophobic cavity of YPs by the passive diffusion of the payload through the porous cell walls in an aqueous suspension of YPs. High terpene loading (1:1 *w*/*w* terpene:YP for materials reported in this publication and up to 3:1 *w*/*w* (unpublished results)) is achieved with this method. Terpene release from YPs is based on passive diffusion out of the particles and is a function of terpene water solubility with complete terpene release in minutes to a few hours [22].

A second-generation approach was developed to better control terpene release from YP encapsulated pro-terpenes (Figure 2b). The pro-terpenes are highly soluble in dimethylsulfoxide (DMSO), a suitable solvent for payload loading in YPs that is removed by lyophilization without loss of pro-terpene. Encapsulation of terpene and pro-terpenes in YPs was assessed qualitatively by microscopy and quantitatively by HPLC. Nile red dye was used to stain terpenes and pro-terpenes to visualize payload encapsulation, as shown in Figure 2c for YP samples containing carvacrol. The HPLC results showed that terpenes and pro-terpenes were encapsulated with >95% efficiency at a target loading ratio of 1:1 terpene:YP or 1.85–1.90:1 pro-terpene:YP (to yield a 1:1 terpene:YP and 0.85–0.9:1 EDTA:YP ratio).

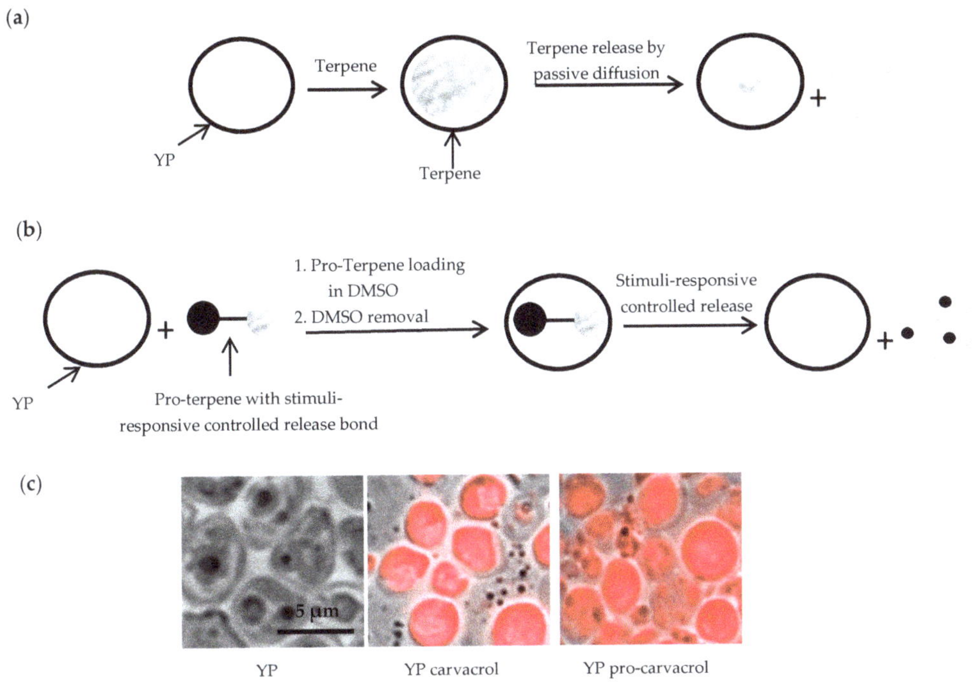

Figure 2. Schematics of (**a**) diffusion-controlled terpene loading in YPs and terpene release, (**b**) pro-terpene loading in YPs and stimuli-controlled terpene release and (**c**) microscopy images of Nile red stained YP control and YPs loaded with carvacrol and pro-carvacrol.

3.1.3. Terpene Release from YP Terpene and YP Pro-Terpene Samples

Terpene release from YP terpene is based on passive diffusion out of the particles and is a function of terpene water solubility. The four terpenes evaluated in this study have similar water solubility (~1 mg/mL) and terpenes are rapidly released from YPs (e.g., >90% carvacrol released at 1 h) upon dilution at concentrations equal or less than 1 mg/mL [22].

The pro-terpenes are water insoluble and their release from YPs is dependent on pH or enzymatic hydrolysis of the ester bonds. We previously showed [23] that the kinetics of pro-terpene hydrolysis is pH dependent with the ester bonds of the pro-terpenes being susceptible to fast hydrolysis in basic pH (>50% pro-carvacrol hydrolyzed within 30 min at pH 10) and increased stability at lower pH values (it took 4 days at pH 7, 21 days at pH 5, and more than 2 months at pH 1.5 to reach 50% hydrolysis of pro-carvacrol) [23].

The YP terpene and YP pro-terpene compositions were evaluated for terpene release in phosphate buffer saline (PBS, pH 7) at 37 °C. The results in Figure 3 clearly show that YP pro-terpenes have greater pH 7 stability than YP terpenes. All YP terpene samples rapidly released their payload in PBS upon dilution of the sample to 1 mg terpene/mL. YP pro-terpenes resuspended in PBS at the same concentration showed a small burst release (<20%), followed by slow hydrolysis and sustained terpene release over two weeks to achieve complete control, and sustained terpene release from YPs.

Next, we evaluated YP terpenes and YP pro-terpenes samples for stability during simulated digestion (Figure 4). Samples were first incubated in simulated gastric fluid (SGF) containing pepsin, followed by incubation in simulated intestinal fluid (SIF) containing pancreatin.

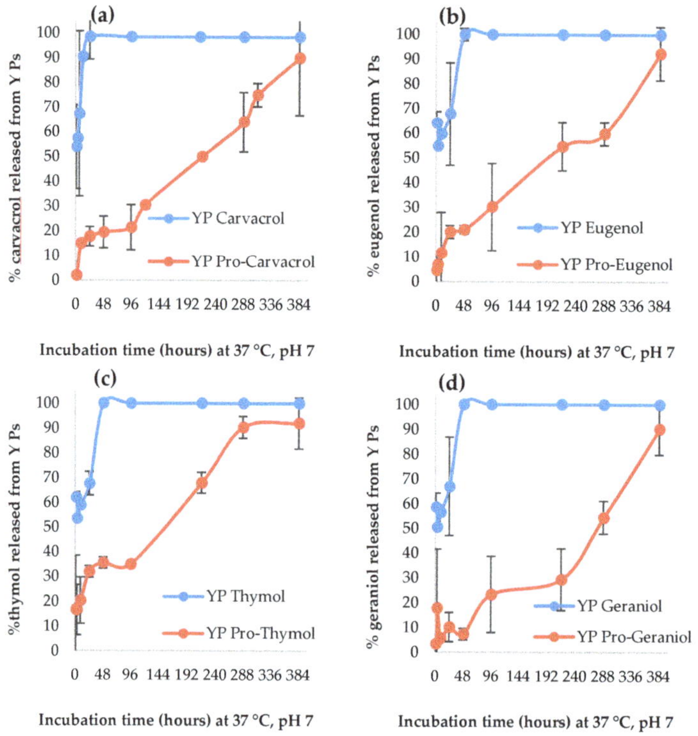

Figure 3. Kinetics of terpene release from YP terpene and YP pro-terpene containing (**a**) carvacrol, (**b**) eugenol, (**c**) thymol, and (**d**) geraniol. Samples were incubated in 0.1 M phosphate buffer saline (PBS, pH 7) at 37 °C at a concentration of 1 mg terpene/mL.

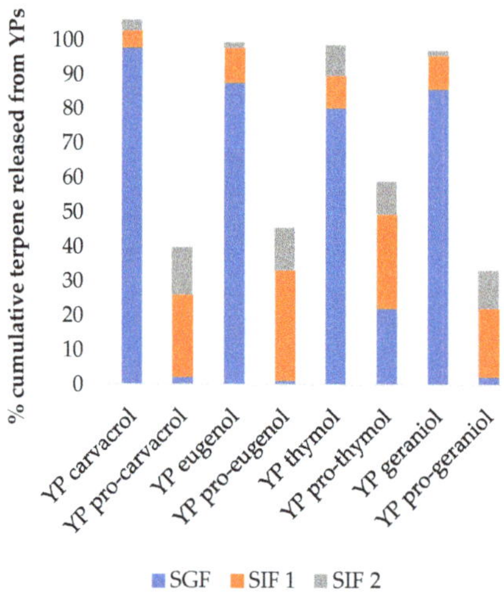

Figure 4. Cumulative terpene released from YPs after 1 h incubation in simulated gastric fluid (SGF) containing pepsin and after two 1 h incubations in fresh simulated intestinal fluid (SIF) with pancreatin.

Terpenes are released from YP terpenes in SGF, primarily due to terpene diffusion out of the particles. YP pro-terpenes are stable in SGF due the pro-terpenes water insolubility, slow hydrolysis at pH < 2, and non-susceptibility to degradation by pepsin (pepsin is an endopeptidase that breaks amide peptide bonds). Terpenes were partially released from YP pro-terpenes in SIF due to the presence of lipases in pancreatin that hydrolyze ester linkages. Pancreatin activity in simulated digestion is lost within 1 h. To demonstrate the effect of enzyme activity on YP pro-terpene hydrolysis and release, samples were subjected to two sequential 1 h incubations in fresh SIF + pancreatin.

3.2. Biological Activity of YP Pro-Terpenes

The biological activity of YP pro-terpenes was evaluated against different model organisms to demonstrate that YP encapsulated pro-terpenes retain the broad-spectrum anti-pathogen effects (antibacterial, antifungal, and anthelmintic) of free terpenes and YP terpenes.

Antimicrobial activity of YP pro-terpenes: YP pro-terpene samples were tested for antibacterial activity against *E. coli*. Table 2 shows the minimum inhibitory concentrations (MICs) of YP samples and controls. Empty YPs have no antimicrobial effect on *E. coli*. For the three terpenes evaluated on *E. coli*, the free terpene, YP terpene and YP pro-terpene samples show similar antibacterial activity. Hydrolysis of pro-terpenes generates EDTA as byproduct and a control sample of YP+EDTA containing the same amount of EDTA generated from pro-terpene hydrolysis was evaluated to confirm that the antimicrobial effect of YP pro-terpene samples was due to the terpene released from the particles and not from the EDTA byproduct.

Table 2. In vitro antibacterial activity on *E. coli* of free terpene, YP terpene and YP pro-terpene samples.

Sample		**MIC 75% (Average of $n = 4$)**
Negative controls	Empty YP	Not active
	YP + EDTA	4244 ± 1400 μg EDTA/mL
Carvacrol samples	Carvacrol	625 ± 0 μg carvacrol/mL
	YP carvacrol	677 ± 45 μg carvacrol/mL
	YP pro-carvacrol	625 ± 0 μg carvacrol/mL
Eugenol samples	Eugenol	938 ± 361 μg eugenol/mL
	YP eugenol	312 ± 0 μg eugenol/mL
	YP pro-eugenol	625 ± 442 μg eugenol/mL
Geraniol samples	Geraniol	1094 ± 313 μg geraniol/mL
	YP geraniol	781 ± 312 μg geraniol/mL
	YP pro-geraniol	1094 ± 312 μg geraniol/mL

The antibacterial effect against *E. coli* is due to bacterial absorption of terpene present in the LB medium added as free terpene or terpene released from YP terpenes (all MICs are below the maximum solubility of terpenes in water). The kinetics of pro-terpene hydrolysis at neutral pH (pH of LB = 6.8) make it unlikely to achieve high enough terpene release from YP pro-terpene samples during the 18 h incubation required for this experiment. Additionally, the pro-terpenes are insoluble in LB medium and YPs are not internalized by *E. coli*; therefore, it is not possible for pro-terpene hydrolysis to occur inside the bacteria. We hypothesized that YP pro-terpenes are hydrolyzed by esterases secreted from *E. coli* into the LB medium. Samples of YP pro-carvacrol were incubated for 18 h in fresh LB or in LB bacterial cell-free spent media collected after *E. coli* culture (LB spent media). The samples were centrifuged to collect the supernatant and YP pellet fractions; carvacrol generated from pro-terpene hydrolysis was quantified in supernatants by HPLC, and the samples were added to *E. coli* to evaluate the antibacterial activity of each fraction (Table 3). The YP pro-carvacrol samples incubated in fresh LB medium released <5% of carvacrol. Samples incubated in spent media released 35 ± 6% carvacrol into the supernatant confirming that pro-terpenes are hydrolyzed by enzymes secreted from *E. coli* into the LB medium. The HPLC analysis of the pellet fractions showed that carvacrol retained in the pellet was in the form of pro-carvacrol for both samples incubated in fresh or spent media. The HPLC results were used to correct carvacrol concentration in supernatant and pellet fractions added to *E. coli*. The results in Table 3 show that only the pellet fraction was active for the sample incubated in fresh LB medium and both fractions collected from spent LB medium were active.

Table 3. In vitro antibacterial activity on *E. coli* of YP pro-carvacrol sample before and after incubation in LB spent media.

LB	Fraction	MIC 75% (μg Carvacrol/mL, Average of $n = 3$)
Fresh	Supernatant	Not active
	Pellet	521 ± 180
Spent	Supernatant	500 ± 216
	Pellet	729 ± 252

Next, we evaluated YP pro-terpenes for antifungal activity against *S. cerevisae*. The results in Table 4 show that empty YP and YP+EDTA are not active against *S. cerevisae* and the three terpenes evaluated in this assay show similar MIC values for free terpene, YP terpene, and YP pro-terpene samples. The activity of YP pro-terpene sample is likely due to a similar effect shown with *E. coli*, with terpene release from YPs upon hydrolysis of pro-terpene induced by esterases secreted by fungi.

Table 4. In vitro antifungal activity on *Saccharomyces cerevisae* of free terpene, YP terpene and YP pro-terpene samples.

Sample		MIC 75% (Average of $n = 3$)
Negative controls	Empty YP	Not active
	YP + EDTA	Not active
Eugenol samples	Eugenol	703 ± 773 g eugenol/mL
	YP eugenol	390 ± 193 μg eugenol/mL
	YP pro-eugenol	781 ± 362 μg eugenol/mL
Thymol samples	Thymol	312 ± 110 μg thymol/mL
	YP thymol	1250 ± 442 μg thymol/mL
	YP pro-thymol	781 ± 312 μg thymol/mL
Geraniol samples	Geraniol	703 ± 773 μg geraniol/mL
	YP geraniol	312 ± 221 μg geraniol/mL
	YP pro-geraniol	260 ± 90 μg geraniol/mL

Anthelmintic activity of YP pro-terpenes: We recently reported the testing of 17 YP terpenes as broad-acting anthelmintics [22]. YP terpenes or subsets of them were active against hookworms (*Ancylostoma ceylanicum* and *Nippostrongylus brasiliensis*) and whipworm (*Trichuris muria*), and overcame albendazole-resistant *Caenorhabditis elegans*. YP encapsulation provides an approach that could lead to the development of anthelmintic terpene formulations for oral delivery. The new YP encapsulation approach using pro-terpenes with stimuli-controlled terpene release could provide materials that overcome fast terpene release and absorption in the stomach.

First, we evaluated if YP pro-terpenes are susceptible to hydrolysis by enzymes in a nematode (*Ascaris suum*) extract. YP pro-carvacrol samples were incubated in *Ascaris* extract at pH 7 and pH 5. Control samples were incubated in buffer only including a control in 0.1 M carbonate buffer (pH 10) for complete pro-carvacrol hydrolysis and carvacrol release from YP. The samples were centrifuged to collect the supernatant and YP pellet fractions, carvacrol was quantified in supernatants, and both fractions were evaluated for antimicrobial activity on *E. coli*. The results in Table 5 show that YP pro-carvacrol was susceptible to hydrolysis by esterases in the *Ascaris* extract with 5–6-fold higher carvacrol release in the *Ascaris* extract compared to buffer controls. There was antimicrobial activity from terpene present in both supernatant and pellet fractions of samples incubated in *Ascaris* extract, but only the pellet fraction of samples incubated in pH 7 and pH 5 buffers is active. The control sample in pH 10 buffer (no worm extract) showed the expected complete carvacrol release and antibacterial activity only in the supernatant fraction.

Table 5. Hydrolysis of YP pro-carvacrol in *Ascaris suum* extract and antibacterial activity on the *E. coli* of the supernatant containing carvacrol released from the YPs and YP pellet fraction containing residual encapsulated pro-carvacrol.

Buffer ± *Ascaris* Extract	% Carvacrol Released from YPs after 24 h Incubation	MIC 75% (µg carvacrol/mL, Average of n = 3)	
		Supernatant	Pellet
PBS (pH 7)	4.5 ± 1.8	Not active	597
Ascaris extract in PBS (pH 7)	27.9 ± 6.0	291 ± 101	451 ± 0
Acetate buffer (pH 5)	6.5 ± 0.9	325	584
Ascaris extract in acetate buffer (pH 5)	37.3 ± 9.5	155 ± 67	261 ± 113
Carbonate buffer (pH 10)	95.6 ± 1.2	598	Not active

Next, the biological activity of YP pro-terpenes and YP terpenes was assessed in three in vitro nematode assays: (1) development of the horse parasite cyathostomin from egg to third-stage larvae [30], (2) toxicity in adult hookworm (*A. ceylanicum*) and (3) toxicity in adult whipworm (*T. muris*).

YP pro-carvacrol and YP carvacrol were evaluated in the cyathostomin E2L assay with samples showing similar dose-response activity (Figure 5) completely inhibiting E2L development at a concentration of 100 µg carvacrol/mL. A control of YP+EDTA also had some toxicity (~40–45% inhibition of larvae development, data not shown) at the highest EDTA concentration equivalent to the expected amount of EDTA generated from the hydrolysis of YP pro-carvacrol at 100 µg/mL. The long incubation period (seven days) required for the Cyathostomin E2L assay likely increased the impact of EDTA complexation of metal ions critical in processes of larvae development.

YP pro-terpene and YP terpene samples were evaluated at a concentration of 333 µg terpene/mL in the adult hookworm and whipworm assays. We previously identified all four terpenes to have a fast-acting effect on hookworm at the selected terpene concentration of 333 µg /mL [22]. We also demonstrated that hookworms readily ingest YPs which leads to two possible mechanisms of terpene entry: (1) hookworm ingestion of terpene released from YPs or (2) hookworm ingestion of YP terpene or YP pro-terpene and subsequent terpene release inside the worms. The results in Figure 6 show that all samples were active against hookworm after 2-h and 24 h incubation and YP+EDTA control was non-toxic.

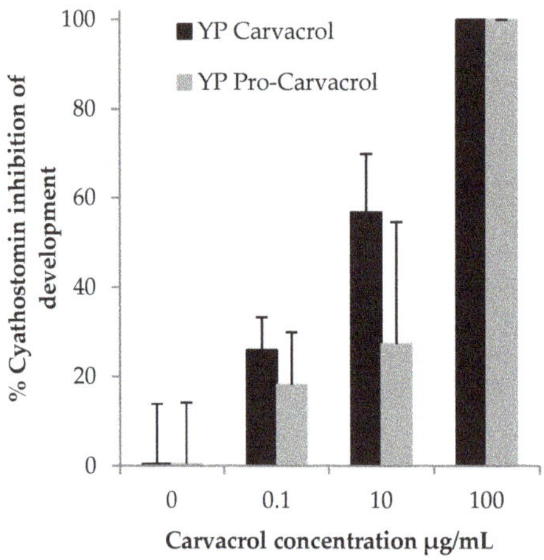

Figure 5. In vitro activity of carvacrol samples in cyathostomin egg-to-larvae (E2L) assay.

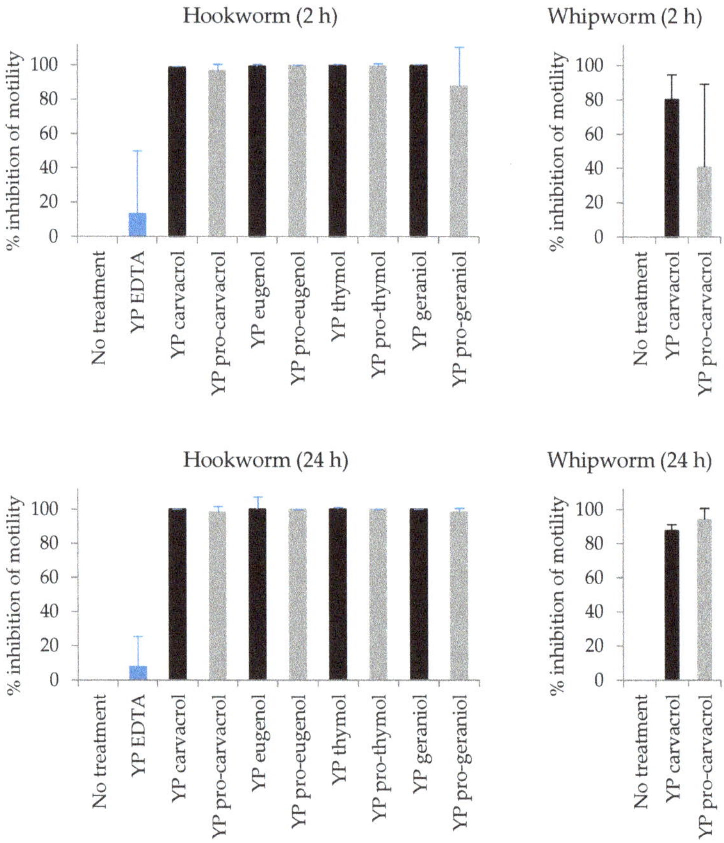

Figure 6. Inhibition of adult hookworm and whipworm motility by YP terpenes and YP pro-terpenes after 2 and 24 h incubation (results are average of eight worms per treatment). Percent inhibition of motility was calculated relative to the mean motility units of control worms (no treatment).

YP pro-carvacrol and YP carvacrol samples were evaluated for in vitro activity in the whipworm assay at a concentration of 333 µg/mL. Unlike hookworms, whipworms do not ingest YPs and therefore toxicity of YP terpenes in whipworm only occurs due to whipworm absorption of terpene released from YPs into the whipworm assay media [22]. The results in Figure 6 show >70% inhibition of whipworm motility after 2 h incubation with YP carvacrol due to fast release of carvacrol from YP carvacrol, but the YP pro-carvacrol sample is less active after 2 h incubation. This reduced activity after 2-h is expected as carvacrol release from YP pro-carvacrol requires hydrolysis of pro-carvacrol mediated by esterases secreted by whipworm into the media, followed by carvacrol diffusion from YPs. Both YP carvacrol and YP pro-carvacrol showed similar inhibition of whipworm motility after 24 h incubation.

4. Discussion

Terpenes are natural products of great commercial interest due to their wide array of functional properties. Microencapsulation of terpenes in some products is challenging due to their high volatility and susceptibility to degradation. We previously developed methods to use yeast particles for the encapsulation of terpenes without the need for alcohols or surfactants. These YP terpenes have been shown to exhibit broad anthelmintic activity [22] and two YP terpene-based products have been developed and commercialized as a fungicide and nematicide for agricultural applications [16–21]. The goals of developing a second generation of YP terpene materials were to improve terpene encapsulation stability and to provide for stimuli-responsive controlled terpene release. This new approach employs pro-terpene compounds that are (1) solid at room temperature to avoid loss due to the high volatility of terpenes, (2) water-insoluble to prevent premature release from YPs, (3) stable at neutral pH, and (4) contain a stimuli-controlled (pH, enzyme) biodegradable bond for controlled terpene release.

High encapsulation efficiency (>95%) of both pro-terpenes or terpenes in YPs was achieved at a target terpene:YP weight ratio of 1:1. Terpene release from YP terpenes is dependent on the diffusion of terpene from the particles and is a function of terpene solubility with complete release in minutes to a few hours upon dilution of YP terpene below its maximum solubility in water. Terpene release from YP pro-terpenes is a function of pro-terpene hydrolysis in response to an external stimulus, followed by terpene diffusion from YPs. Terpene encapsulation stability is improved using pro-terpenes, extending the release of terpene at pH 7 from a few hours (YP terpene) up to two weeks (YP pro-terpenes).

The YP pro-terpene and YP terpene samples showed similar biological activity in antibacterial, antifungal and anthelmintic in vitro assays. The activity of YP pro-terpene in these assays is mediated by the presence of esterases to induce hydrolysis of the pro-terpene and subsequent terpene release from YPs. The enhanced stability and controlled release of YP pro-terpenes could allow for developing terpene formulations for applications such as (1) environmental biocontrol agent applied directly to contaminated soil or a pass through in feed to reduce transmission of gastrointestinal parasitic nematodes by targeting developing/infectious larvae of parasites that infect livestock and humans in the third larval stage (e.g., cyathostomins, *Haemonchus*, *Ostertagia*, hookworms), (2) formulations for oral delivery of terpenes for targeting of internal gastrointestinal parasites, and (3) controlled-release of terpenes as food fragances and flavors, or as antimicrobial agents in food preservation. Future work will focus on the in vivo evaluation of YP pro-terpenes.

This new approach of payload encapsulation in YPs using pro-terpenes can be expanded to a broad range of small drug molecules that are difficult to trap using previously developed yeast particle drug encapsulation methods. We are currently investigating this approach to stably generate encapsulated yeast particle pro-drugs with stimuli-controlled drug release for a broad range of payload molecules such as antibacterials (e.g., isoniazid, oxazolidinones, and cycloserine), anti-inflammatories (e.g., naproxen, ibuprofen), and chemotherapeutics (e.g., doxorubicin).

5. Conclusions

Yeast particles can be used for the encapsulation of scaffolded terpene compounds containing a biodegradable linker to control terpene release. These YP pro-terpene samples exhibit the same loading capacity as YP terpenes, and with additional benefits of improved stability and control over terpene release in response to pH or esterase induced hydrolysis of the pro-terpene compound. Both YP terpenes and YP pro-terpenes exhibit biological activity in antimicrobial and anthelmintic assays.

6. Patents

Yeast Cell Wall Particle Encapsulation of Biodegradable Pro-Payloads. G.R. Ostroff and E. R. Soto. US Patent App. 16/981,072, 2021. 28 January 2021.

Author Contributions: Conceptualization, E.R.S. and G.R.O.; methodology, E.R.S., F.R., H.L., C.G., J.C., M.E., D.G., M.K.N. and J.F.U.J.; formal analysis, E.R.S. and G.R.O.; resources, R.V.A. and G.R.O.; writing—original draft preparation, E.R.S.; writing—review and editing, R.V.A. and G.R.O.; project administration, G.R.O.; funding acquisition, R.V.A. and G.R.O. All authors have read and agreed to the published version of the manuscript.

Funding: This research was funded by (1) the National Institutes of Health/National Institute of Allergy and Infectious Diseases grants R01AI056189 and R01AI50866 to R.V.A., (2) Agriculture and Food Research Initiative Competitive Grant no. 2015–11323 from the USDA National Institute of Food and Agriculture to R.V.A., and (3) Sponsored Research Agreement from Eden Research plc to G.R.O.

Institutional Review Board Statement: All animal experiments were carried out under protocols approved by the Institutional Animal Care and Use Committee (IACUC) of the University of Massachusetts Medical School protocols #A-2483 and #A-2484, the University of Kentucky protocol #2015–2078, and the USDA/ARS/Beltsville #17–019 and #20–003. Hamsters and mice used in this study were housed, handled, fed and experimentally used following the National Institute of Health (NIH) Guide for the Care and Use of Laboratory Animals in Research. Euthanasia was performed by CO_2 asphyxiation, followed by bilateral pneumothorax.

Data Availability Statement: The data presented in this study are available upon request from the corresponding author.

Conflicts of Interest: G.R.O. and E.R.S. are inventors of US Patent App. 16/981,072, 2021. Yeast Cell Wall Particle Encapsulation of Biodegradable Pro-Payloads assigned to University of Massachusetts Medical School. The patent is managed by the Office of Technology Management and the Conflicts of Interest policies of the University of Massachusetts. The other authors declare no conflict of interest.

References

1. Cowan, M.M. Plant products as antimicrobial agents. *Clin. Microbiol. Rev.* **1999**, *12*, 564–582. [CrossRef]
2. Falleh, H.; Ben Jemaa, M.; Saada, M.; Ksouri, R. Essential oils: A promising eco-friendly food preservative. *Food Chem.* **2020**, *330*, 127268. [CrossRef]
3. Ben Salha, G.; Abderrabba, M.; Labidi, J. A status review of terpenes and their separation methods. *Rev. Chem. Eng.* **2019**, *37*, 433–447. [CrossRef]
4. Bakry, A.M.; Abbas, S.; Ali, B.; Majeed, H.; Abouelwafa, M.Y.; Mousa, A.; Liang, L. Microencapsulation of Oils: A Comprehensive Review of Benefits, Techniques, and Applications. *Compr. Rev. Food Sci. Food Saf.* **2016**, *15*, 143–182. [CrossRef]
5. Calo, J.R.; Crandall, P.G.; O'Bryan, C.A.; Ricke, S.C. Essential oils as antimicrobials in food systems—A review. *Food Control* **2015**, *54*, 111–119. [CrossRef]
6. Saifullah, M.; Shishir, M.R.I.; Ferdowsi, R.; Tanver Rahman, M.R.; Van Vuong, Q. Micro and nano encapsulation, retention and controlled release of flavor and aroma compounds: A critical review. *Trends Food Sci. Technol.* **2019**, *86*, 230–251. [CrossRef]
7. Bhalerao, Y.P.; Wagh, S.J. A Review on thymol encapsulation and its controlled release through biodegradable polymer shells. *Int. J. Pharm. Sci. Res.* **2018**, *9*, 4522–4532.
8. De Matos, S.P.; Teixeira, H.F.; De Lima, A.A.N.; Veiga-Junior, V.F.; Koester, L.S. Essential oils and isolated terpenes in nanosystems designed for topical administration: A review. *Biomolecules* **2019**, *9*, 138. [CrossRef] [PubMed]
9. Soto, E.R.; Ostroff, G.R. Characterization of multilayered nanoparticles encapsulated in yeast cell wall particles for DNA. *Bioconjug. Chem.* **2008**, *19*, 840–848. [CrossRef] [PubMed]
10. Aouadi, M.; Tesz, G.J.; Nicoloro, S.M.; Wang, M.; Chouinard, M.; Soto, E.; Ostroff, G.R.; Czech, M.P. Orally delivered siRNA targeting macrophage Map4k4 suppresses systemic inflammation. *Nature* **2009**, *458*, 1180–1184. [CrossRef]

11. Soto, E.; Ostroff, G. Glucan particles as carriers of nanoparticles for macrophage-targeted delivery. In *Nanomaterials for Biomedicine*; American Chemical Society: Washington, DC, USA, 2012; pp. 57–79. [CrossRef]
12. Mirza, Z.; Soto, E.R.; Dikengil, F.; Levitz, S.M.; Ostroff, G.R. Beta-glucan particles as vaccine adjuvant carriers. In *Vaccines for Invasive Fungal Infections*; Humana Press: Totowa, NJ, USA, 2017; pp. 143–157.
13. Soto, E.R.; O'Connell, O.; Dikengil, F.; Peters, P.J.; Clapham, P.R.; Ostroff, G.R. Targeted Delivery of Glucan Particle Encapsulated Gallium Nanoparticles Inhibits HIV Growth in Human Macrophages. *J. Drug Deliv.* **2016**, *2016*, 8520629. [CrossRef]
14. Hamza, Z.; El-Hashash, M.; Aly, S.; Hathout, A.; Soto, E.; Sabry, B.; Ostroff, G. Preparation and characterization of yeast cell wall beta-glucan encapsulated humic acid nanoparticles as an enhanced aflatoxin B1 binder. *Carbohydr. Polym.* **2019**, *203*, 185–192. [CrossRef] [PubMed]
15. Huang, H.; Ostroff, G.R.; Lee, C.K.; Specht, C.A.; Levitz, S.M. Robust Stimulation of Humoral and Cellular Immune Responses following Vaccination with Antigen-Loaded β-Glucan Particles. *MBio.* **2010**, *1*, e00164-10. [CrossRef]
16. Franklin, L.; Ostroff, G.; Harman, G. Terpene-Containing Compositions and Methods of Making and Using Them. U.S. Patent No. 10,667,512, 2 June 2020.
17. Franklin, L.; Ostroff, G. Compositions and Methods Comprising Terpenes or Terpene Mixtures Selected from Thymol, Eugenol, Geraniol, Citral and L-Carvone. U.S. Patent No. 10,258,033, 16 April 2019.
18. Franklin, L.; Ostroff, G. Compositions Containing a Hollow Glucan Particle or a Cell Wall Particle Encapsulating a Terpene Component, Methods of Making and Using Them. U.S. Patent No. 10,638,750, 5 May 2020.
19. Franklin, L.; Ostroff, G. Nematicidal Compositions and Methods of Using Them. U.S. Patent No. 10,004,229, 26 June 2018.
20. Franklin, L.; Ostroff, G. Nematicidal Compositions and Methods of Using Them. U.S. Patent No. 9,655,360, 23 May 2017.
21. Franklin, L.; Ostroff, G. Compositions and Methods Comprising Terpenes or Terpene Mixtures Selected from Thymol, Eugenol, Geraniol, Citral, and L-Carvone. U.S. Patent No. 9,439,416, 13 September 2016.
22. Mirza, Z.; Soto, E.R.; Hu, Y.; Nguyen, T.T.; Koch, D.; Aroian, R.V.; Ostroff, G.R. Anthelmintic activity of yeast particle-encapsulated terpenes. *Molecules* **2020**, *25*, 2958. [CrossRef] [PubMed]
23. Soto, E.; Rus, F.; Ostroff, G.R. Yeast Cell Wall Particle Encapsulation of Pro-Terpene Payloads. Nanotech 2019 TechConnect Briefs. 2019. Available online: https://briefs.techconnect.org/papers/yeast-cell-wall-particle-encapsulation-of-pro-terpene-payloads/ (accessed on 10 January 2021).
24. Carbone-Howell, A.L.; Stebbins, N.D.; Uhrich, K.E. Poly(anhydride-esters) comprised exclusively of naturally occurring antimicrobials and EDTA: Antioxidant and antibacterial activities. *Biomacromolecules* **2014**, *15*, 1889–1895. [CrossRef] [PubMed]
25. Sultanbawa, Y.; Cusack, A.; Currie, M.; Davis, C. An innovative microplate assay to facilitate the detection of antimicrobial activity in plant extracts. *J. Rapid Methods Autom. Microbiol.* **2009**, *17*, 519–534. [CrossRef]
26. Harley, C.A.; Tipper, D.J. The role of charged residues in determining transmembrane protein insertion orientation in yeast. *J. Biol. Chem.* **1996**, *271*, 24625–24633. [CrossRef]
27. U.S. Pharmacopeia. *Simulated Gastric Fluid, TS. National Formulary No. 24/19*; National Publishing: Philadelphia, PA, USA, 2000.
28. U.S. Pharmacopeia. *Simulated Intestinal Fluid*; United States Pharmacopeial Convention: Rockville, MD, USA, 1995.
29. Urban, J.F., Jr.; Hu, Y.; Miller, M.M.; Scheib, U.; Yiu, Y.Y.; Aroian, R.V. *Bacillus thuringiensis*-derived Cry5B has potent anthelmintic activity against *Ascaris suum*. *PLoS Negl. Trop. Dis.* **2013**, *7*, e2263. [CrossRef]
30. Hu, Y.; Miller, M.; Zhang, B.; Nguyen, T.T.; Nielsen, M.K.; Aroian, R.V. In vivo and in vitro studies of Cry5B and nicotinic acetylcholine receptor agonist anthelmintics reveal a powerful and unique combination therapy against intestinal nematode parasites. *PLoS Negl. Trop. Dis.* **2018**, *12*, e0006506. [CrossRef]
31. Lyons, E.T.; Drudge, J.H.; Tolliver, S.C. Prevalence of some internal parasites found (1971-1989) in horses born on a farm in Central Kentucky. *J. Equine Vet. Sci.* **1990**, *10*, 99–107. [CrossRef]
32. Hu, Y.; Zhan, B.; Keegan, B.; Yiu, Y.Y.; Miller, M.M.; Jones, K.; Aroian, R.V. Mechanistic and Single-Dose In Vivo Therapeutic Studies of Cry5B Anthelmintic Action against Hookworms. *PLoS Negl. Trop. Dis.* **2012**, *6*, e1900. [CrossRef] [PubMed]
33. Hu, Y.; Xiao, S.H.; Aroian, R.V. The new anthelmintic tribendimidine is an L-type (Levamisole and Pyrantel) nicotinic acetylcholine receptor agonist. *PLoS Negl. Trop. Dis.* **2009**, *3*, e499. [CrossRef] [PubMed]
34. Marcellino, C.; Gut, J.; Lim, K.C.; Singh, R.; McKerrow, J.; Sakanari, J. WormAssay: A novel computer application for whole-plate motion-based screening of macroscopic parasites. *PLoS Negl. Trop. Dis.* **2012**, *6*, e1494. [CrossRef] [PubMed]
35. PubChem. Available online: https://pubchem.ncbi.nlm.nih.gov/ (accessed on 17 December 2020).